T0281403

SECOND EDITION

Harmonics, Power Systems, and Smart Grids

SECOND EDITION

Harmonics, Power Systems, and Smart Grids

FRANCISCO C. DE LA ROSA
CONSULTANT, CONROE, TEXAS, USA

CRC Press
Taylor & Francis Group
Boca Raton London New York

CRC Press is an imprint of the
Taylor & Francis Group, an **informa** business

CRC Press
Taylor & Francis Group
6000 Broken Sound Parkway NW, Suite 300
Boca Raton, FL 33487-2742

First issued in paperback 2020

© 2015 by Taylor & Francis Group, LLC
CRC Press is an imprint of Taylor & Francis Group, an Informa business

No claim to original U.S. Government works

ISBN-13: 978-1-4822-4383-3 (hbk)
ISBN-13: 978-0-367-65604-1 (pbk)

Visit the Taylor & Francis Web site at
http://www.taylorandfrancis.com

and the CRC Press Web site at
http://www.crcpress.com

To the memory of my departed loved ones

To my beloved mother, wife, and son

Contents

Preface

This book seeks to provide a comprehensive reference on harmonic current generation, propagation, and control in electrical power networks, including the broadly cited smart grid. Harmonic waveform distortion is one of the most important issues that the electric industry faces today due to the substantial volume of electric power that is converted from alternating current (AC) to other forms of electricity required in multiple applications. It is also a topic of much discussion in technical working groups that issue recommendations and standards for waveform distortion limits. Equipment manufacturers and electric utilities strive to find the right conditions to design and operate power apparatuses that can reliably operate in harmonic environments and, at the same time, meet harmonic emission levels within recommended values.

This book provides a compilation of the most relevant aspects of harmonics in a way that the unfamiliar reader can grasp and use to get up to date with the subject matter and that the experienced reader can use to navigate directly to specific subjects of interest.

An introductory description on the definition of harmonics along with analytical expressions for electrical parameters under nonsinusoidal situations is provided in Chapter 1. This is followed in Chapter 2 by descriptions of the different sources of harmonics that are concerns for the electric industry.

Industrial facilities are by far the major producers of harmonic currents. Most industrial processes involve one form or another of power conversion to run processes that use large direct current (DC) motors or variable frequency drives. Others feed large electric furnaces, electric welders, or battery chargers, which are formidable generators of harmonic currents. How harmonic current producers have spread from industrial to commercial and residential facilities—mostly as a result of the proliferation of personal computers and entertaining devices that require rectified power—is described. Additionally, the use of energy-saving devices, such as electronic ballasts in commercial lighting and interruptible power supplies that provide voltage support during power interruptions, makes the problem even larger.

As this takes place, standards bodies struggle to adapt present regulations on harmonics to levels more in line with realistic scenarios and to avoid compromising the reliable operation of equipment at utilities and customer locations. The most important and widely used industry standards to control harmonic distortion levels are described in Chapter 3.

The effects of harmonics are thoroughly documented in technical literature. They range from accelerated equipment aging to abnormal operation of sensitive processes or protective devices. Chapter 4 summarizes the most relevant effects of harmonics in different situations that equally affect residential, commercial, and industrial customers. A particular effort is devoted to illustrating the effects of harmonics in electrical machines related to pulsating torques that can drive machines into excessive shaft vibration.

Given the extensive dissemination of harmonic sources in the electrical network, monitoring harmonic distortion at the interface between customer and supplier has become essential. Additionally, the dynamics of industrial loads require the characterization of harmonic distortion levels over extended periods. Chapter 5 summarizes the most relevant aspects and industry recommendations to take into account when deciding to undertake the task of characterizing harmonic levels at a given facility.

One of the most effective methods to mitigate the effect of harmonics is the use of passive filters. Chapter 6 provides a detailed description of their operation principle and design. Single-tuned and high-pass filters are included in this endeavor. Simple equations that involve the AC source data, along with the parameters of other important components (particularly the harmonic-generating source), are described. Filter components are determined and tested to meet industry standards' operation performance. Some practical examples are used to illustrate the application of the different filtering schemes. The Active Filter Concept, which represents a more sophisticated option for harmonic control, is briefly described.

Because of the expenses incurred in providing harmonic filters, particularly but not exclusively at industrial installations, other methods to alleviate the harmonic distortion problem are often applied. Alternative methods, including use of stiffer AC sources, power converters with increased number of pulses, series reactors, and load reconfiguration, are presented in Chapter 7.

In Chapter 8, a description of the most relevant elements that play a role in the study of the propagation of harmonic currents in a distribution network is presented. These elements include the AC source, transmission lines, cables, transformers, harmonic filters, power factor, capacitor banks, etc. In dealing with the propagation of harmonic currents in electrical networks, it is very important to recognize the complexity reached when extensive networks are considered. Therefore, some examples are illustrated to show the convenience of using specialized tools in the analysis of complicated networks with multiple harmonic sources. The penetration of harmonic currents in the electrical network that can affect adjacent customers and even reach the substation transformer is also discussed.

The most important aspects to determine power losses in electrical equipment attributed to harmonic waveform distortion are presented in Chapter 9. This is done with particular emphasis on transformers and rotating machines.

Most of the examples presented in this book are based on the author's experience in industrial applications, for the most part in the petroleum and electric power utility environments.

A useful addition to the revised edition of the book opens with the smart grid concept discussion in Chapter 10, which portrays a state of affairs in the development, testing, and integration of new devices in the electric power grid, intended to become part of the smart grid. The portrayal includes a depiction of the multiple players in this endeavor, but focus is maintained in the increased levels of harmonic distortion expected with the growing amounts of solid electronic devices that the smart grid concept entails. Chapter 11 describes the characteristic harmonics in the smart grid world, including those from solar and wind power converters and power electronics in FACTS and HVDC technologies.

Finally, Chapter 12 concludes the book, touching on harmonics from the latest innovative electric grid technologies, which includes electric vehicles connected to the grid, superconductive fault current limiters, and electric vehicle charging stations.

I hope this book provides a useful contribution in identifying the relevant aspects of a complex phenomenon to better devise harmonic control measures in a variety of applications.

Francisco C. De La Rosa
Conroe, Texas

Acknowledgments

My appreciation for the publication of this book goes to my family for their absolute support. Thanks to Connie, my wife, for bearing with me at all times and especially during the period when this book was being written.

To produce some of the computer-generated plots presented in this book, I used a number of software tools that were of utmost importance to illustrate fundamental concepts and application examples. Thanks to Professor Mack Grady at the University of Texas at Austin for allowing me to use his HASIP software and to Tom Grebe at Electrotek Concepts, Inc., for granting me permission to use Electrotek Concepts TOP, The Output Processor. The friendly PSCAD (free) student version from Manitoba HVDC Research Centre, Inc., was instrumental in producing many of the illustrations presented in this book, and a few examples were also generated with the free Power Quality Teaching Toy Tool from Alex McEachern at Power Standards Lab.

About the Author

Francisco C. De La Rosa, lead of the Electric Power Systems team at Mott MacDonald, Inc., in The Woodland, Texas, earned BSc and MSc degrees in industrial and power engineering from Coahuila and Monterrey Technological Institutes in Mexico, respectively, and a PhD degree in electrical engineering from Uppsala University in Sweden.

In his professional experience, Dr. De La Rosa has previously held a number of positions in industry, including senior director of his own office in Houston, Texas, and St. Louis, Missouri; director of electrical engineering at Bruker Energy and Supercon Technologies and at Zenergy Power in California; senior electrical engineer at EPS International and staff scientist at Distribution Control Systems (now Aclara Power Line Systems) in Missouri; and research scientist at Instituto de Investigaciones Eléctricas in Mexico. For more than 20 years, he has conducted research, tutored, and offered engineering consultancy services for electric, oil, steel mill, and electric railway companies in the United States, Canada, Mexico, and Venezuela. Dr. De La Rosa taught electrical engineering courses at the Nuevo Leon State University in Monterrey, Mexico, as an invited lecturer from 2000 to 2001. He is particularly fascinated by smart grid technology and the integration of renewable energy in the electric power grid. He holds professional membership in the IEEE Power Engineering Society, where he participates in working groups dealing with superconducting fault current limiters, harmonics, power quality, the smart grid, and distributed generation. Dr. De La Rosa is also a member of CIGRE.

1

Fundamentals of Harmonic Distortion and Power Quality Indices in Electric Power Systems

1.1 Introduction

Ideally, an electricity supply should invariably deliver a perfectly sinusoidal voltage signal at every customer location. However, for a number of reasons, utilities often find it hard to preserve such desirable conditions. The deviation of the voltage and current waveforms from sinusoidal is described in terms of the waveform distortion, often expressed as harmonic distortion.

Harmonic distortion is not new, and it constitutes at present one of the main concerns for engineers in the several stages of energy utilization within the power industry. In the first electric power systems, harmonic distortion was mainly caused by saturation of transformers, industrial arc furnaces, and other arc devices like large electric welders. The major concern was the effect that harmonic distortion could have on electric machines, telephone interference, and increased risk of faults from overvoltage conditions developed on power factor correction capacitors.

In the past, harmonics represented less of a problem due to the conservative design of power equipment and the common use of delta-grounded wye connections in distribution transformers.

The increasing use of nonlinear loads in industry is keeping harmonic distortion in distribution networks on the rise. The most used nonlinear device is perhaps the static power converter so widely used in industrial applications in the steel, paper, and textile industries. Other applications include multipurpose motor speed control, electrical transportation systems, and electrodomestic appliances. By 2000, it was estimated that electronic loads accounted for around half of U.S. electrical demand, and much of that growth in electronic load involved the residential sector.[1]

A situation that has raised waveform distortion levels in distribution networks even further is the application of capacitor banks used in industrial plants for power factor correction and by power utilities for increasing voltage profile along distribution lines. The resulting reactive impedance forms a tank circuit with the system inductive reactance at a certain frequency likely to coincide with one of the characteristic harmonics of the load. This condition will trigger large oscillatory currents and voltages that may stress the insulation. This situation imposes a serious challenge to industry and utility engineers to pinpoint and correct excessive harmonic waveform distortion levels on the waveforms because its steady increase happens to take place right at the time when the use of sensitive electronic equipment is on the rise.

No doubt harmonic studies from the planning to the design stages of power utility and industrial installations will prove to be an effective way to keep networks and equipment under acceptable operating conditions and to anticipate potential problems with the installation or addition of nonlinear loads.

1.2 Basics of Harmonic Theory

The term *harmonics* originated in the field of acoustics, where it was related to the vibration of a string or an air column at a frequency that is a multiple of the base frequency. A harmonic component in an AC power system is defined as a sinusoidal component of a periodic waveform that has a frequency equal to an integer multiple of the fundamental frequency of the system.

Harmonics in voltage or current waveforms can then be conceived as perfectly sinusoidal components of frequency multiples of the fundamental frequency:

$$f_h = (h) \times (\text{fundamental frequency}) \tag{1.1}$$

where h is an integer.

For example, a fifth harmonic would yield a harmonic component:

$$f_h = (5) \times (60) = 300 \text{ Hz} \quad \text{and} \quad f_h = (5) \times (50) = 250 \text{ Hz}$$

in 60 and 50 Hz systems, respectively.

Figure 1.1 shows an ideal 60 Hz waveform with a peak value of around 100 A, which can be taken as one per unit. Likewise, it also portrays waveforms of amplitudes (1/7), (1/5), and (1/3) per unit and frequencies seven, five, and three times the fundamental frequency, respectively. This behavior showing harmonic components of decreasing amplitude often following an inverse law with harmonic order is typical in power systems.

These waveforms can be expressed as

$$i_1 = Im_1 \sin \omega t \tag{1.2}$$

$$i_3 = Im_3 \sin(3\omega t - \delta_3) \tag{1.3}$$

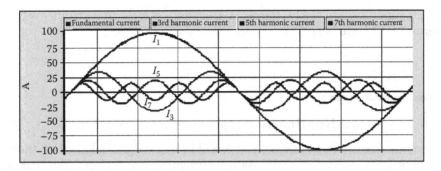

FIGURE 1.1
Sinusoidal 60 Hz waveform and some harmonics.

$$i_5 = Im_5 \sin(5\omega t - \delta_5) \tag{1.4}$$

$$i_7 = Im_7 \sin(7\omega t - \delta_7) \tag{1.5}$$

where Im_h is the peak rms value of the harmonic current h.

Figure 1.2 shows the same harmonic waveforms as those in Figure 1.1 superimposed on the fundamental frequency current yielding I_{total}. If we take only the first three harmonic components, the figure shows what a distorted current waveform at the terminals of a six-pulse converter would look like. All additional harmonics would impose a further waveform distortion.

The resultant distorted waveform can thus be expressed as

$$I_{total} = Im_1 \sin \omega t + Im_3 \sin(3\omega t - \delta_3) + Im_5 \sin(5\omega t - \delta_5) + Im_7 \sin(7\omega t - \delta_7) \tag{1.6}$$

In short, a summation of perfectly sinusoidal waveforms can give rise to a distorted waveform. Conversely, a distorted waveform can be represented

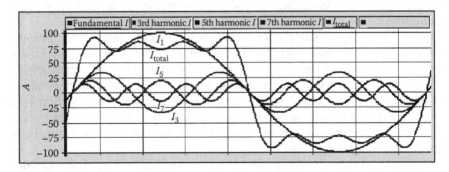

FIGURE 1.2
Sinusoidal waveform distorted by third, fifth, and seventh harmonics.

as the superposition of a fundamental frequency waveform with other waveforms of different harmonic frequencies and amplitudes.

1.3 Linear and Nonlinear Loads

From the discussion in this section, it will be evident that a load that draws current from a sinusoidal AC source presenting a waveform like that of Figure 1.2 cannot be conceived as a linear load.

1.3.1 Linear Loads

Linear loads are those in which voltage and current signals follow one another very closely, such as the voltage drop that develops across a constant resistance, which varies as a direct function of the current that passes through it. This relation is better known as Ohm's law and states that the current through a resistance fed by a varying voltage source is equal to the relation between the voltage and the resistance, as described by

$$i(t) = \frac{v(t)}{R} \tag{1.7}$$

This is why the voltage and current waveforms in electrical circuits with linear loads look alike. Therefore, if the source is a clean open circuit voltage, the current waveform will look identical, showing no distortion. Circuits with linear loads thus make it simple to calculate voltage and current waveforms. Even the amounts of heat created by resistive linear loads like heating elements or incandescent lamps can easily be determined because they are proportional to the square of the current. Alternatively, the involved power can also be determined as the product of the two quantities, voltage and current.

Other linear loads, such as electrical motor driving fans, water pumps, oil pumps, cranes, elevators, etc., not supplied through power conversion devices like variable frequency drives or any other form or rectification/inversion of current will incorporate magnetic core losses that depend on iron and copper physical characteristics. Voltage and current distortion may be produced if ferromagnetic core equipment is operated on the saturation region, a condition that can be reached, for instance, when equipment is operated above rated values.

Capacitor banks used for power factor correction by electric companies and industry are another type of linear load. Figure 1.3 describes a list of linear loads.

Resistive elements	Inductive elements	Capacitive elements

• Incandescent lighting
• Electric heaters
	• Induction motors
	• Current limiting reactors
	• Induction generators
	  (windmills)
	• Damping reactors used
	  to attenuate harmonics
	• Tuning reactors in
	  harmonic filters
		• Power factor correction
		  capacitor banks
		• Underground cables
		• Insulated cables
		• Capacitors used in
		  harmonic filters

FIGURE 1.3
Examples of linear loads.

A voltage and current waveform in a circuit with linear loads will show the two waveforms in phase with one another. Voltage and current involving inductors make voltage lead current, and circuits that contain power factor capacitors make current lead voltage. Therefore, in both cases, the two waveforms will be out of phase from one another. However, no waveform distortion will take place.

Figure 1.4 presents the relation among voltage, current, and power in a linear circuit consisting of an AC source feeding a purely resistive circuit. Notice that instantaneous power, $P = V * I$, is never negative because both waveforms are in phase and their product will always yield a positive quantity. The same result is obtained when power is obtained as the product of the resistance with the square of the current.

Figure 1.5(a) shows the relation between the same parameters for the case when current I lags the voltage V, which would correspond to an inductive load, and Figure 1.5(b) for the case when I leads the voltage V, as in the case of a capacitive load. Negative and positive displacement power factors (discussed in Section 1.5) are related to Figure 1.5(a) and (b), respectively. Note that in these cases the product $V * I$ has positive and negative values. The positive values correspond to the absorption of current by the load and the negative values to the flux of current toward the source.

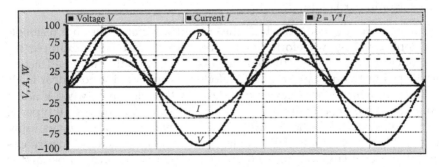

FIGURE 1.4
Relation among voltage, current, and power in a purely resistive circuit.

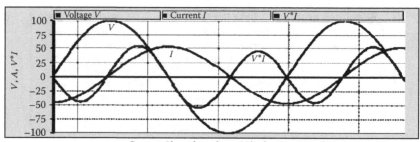

Current *I* lags the voltage *V* (inductive circuit)

(a)

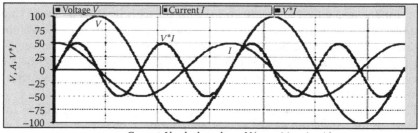

Current *I* leads the voltage *V* (capacitive circuit)

(b)

FIGURE 1.5

Relation among voltage, current, and their product in (a) inductive and (b) capacitive circuits, respectively.

In any case, the sinusoidal nature of voltage and current waveforms is preserved, just as in the case of Figure 1.4 that involves a purely resistive load. Observe that even the product $V * I$ has equal positive and negative cycles with a zero average value; it is positive when V and I have the same polarity and negative when V and I have different polarity.

1.3.2 Nonlinear Loads

Nonlinear loads are loads in which the current waveform does not resemble the applied voltage waveform due to a number of reasons, for example, the use of electronic switches that conduct load current only during a fraction of the power frequency period. Therefore, we can conceive nonlinear loads as those in which Ohm's law cannot describe the relation between V and I. Among the most common nonlinear loads in power systems are all types of rectifying devices, like those found in power converters, power sources, uninterruptible power supply (UPS) units, and arc devices like electric furnaces and fluorescent lamps. Figure 1.6 provides a more extensive list of various devices in this category. As later discussed in Chapter 4, *nonlinear loads* cause a number of disturbances, like voltage waveform distortion, overheating in

Power electronics	ARC devices
• Power converters	• Fluorescent lighting
• Variable frequency drives	• ARC furnaces
• DC motor controllers	• Welding machines
• Cycloconverters	
• Cranes	
• Elevators	
• Steel mills	
• Power supplies	
• UPS	
• Battery chargers	
• Inverters	

FIGURE 1.6
Examples of some nonlinear loads.

transformers and other power devices, overcurrent on equipment-neutral connection leads, telephone interference, and microprocessor control problems, among others. Chapters 11 and 12 describe some additional sources of harmonics in the new smart grid setting and those from the latest electric grid technologies, respectively.

Figure 1.7 shows the voltage and current waveforms during the switching action of an insulated gate bipolar transistor (IGBT), a common power electronics solid-state device. This is the simplest way to illustrate the performance of a nonlinear load in which the current does not follow the sinusoidal source voltage waveform, except during the time when firing pulses FP1 and FT2 (as shown on the lower plot) are ON. Some motor speed controllers, household equipment like TV sets and VCRs, and a large variety of other residential and commercial electronic equipment use this type of voltage control. When the same process takes place in three-phase equipment and the amount of load is significant, a corresponding distortion can also take place in the voltage signal.

Even linear loads like power transformers can act nonlinear under saturation conditions. What this means is that, in certain instances, the magnetic flux density (B) in the transformer ceases to increase or increases very little as the magnetic flux intensity (H) keeps growing. This occurs beyond the so-called saturation knee of the magnetizing curve of the transformer. The behavior of the transformer under changing cycles of positive and negative values of H is shown in Figure 1.8 and is known as hysteresis curve.

Of course, this nonlinear effect will last as long as the saturation condition prevails. For example, an elevated voltage can be fed to the transformer during low load conditions that can last up to several hours, whereas an overloaded transformer condition is often observed during starting of large motors or high inertia loads in industrial environments lasting a few seconds. The same situation can occur practically with other types of magnetic core devices.

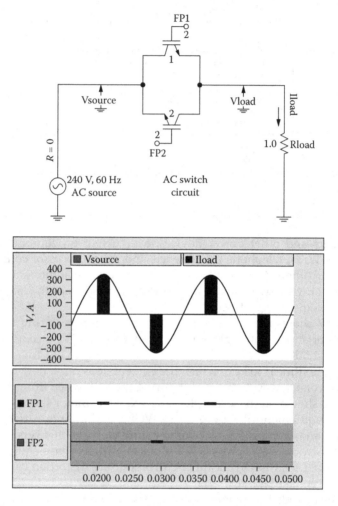

FIGURE 1.7
Relation between voltage and current in a typical nonlinear power source.

In Figure 1.8, the so-called transformer magnetizing curve of the transformer (curve 0–1) starts at point 0 with the increase of the magnetic field intensity H, reaching point 1 at peak H, beyond which the magnetic flux shows a flat behavior, i.e., a small increase in B on a large increase in H. Consequently, the current starts getting distorted and thus showing harmonic components on the voltage waveform too. Notice that from point 1 to point 2, the $B–H$ characteristic follows a different path so that when magnetic field intensity has decreased to zero, a remanent flux density, Br, called *permanent magnetization* or *remanence*, is left in the transformer core. This is canceled only when electric field intensity reverses and reaches the so-called *coercive force Hc*. Point 4 corresponds to the negative cycle magnetic field intensity peak. When H

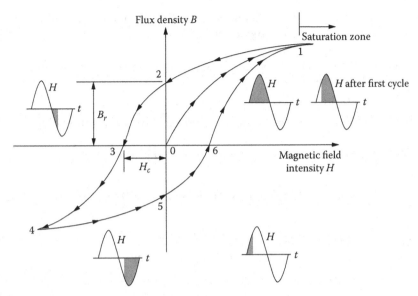

FIGURE 1.8
Transformer hysteresis characteristic.

returns to zero at the end of the first cycle, the *B–H* characteristic ends in point 5. From here a complete hysteresis cycle would be completed when *H* again reaches its peak positive value to return to point 1.

The area encompassed by the hysteresis curve is proportional to the transformer core losses. It is important to note that transformer cores that offer a small coercive force would be needed to minimize losses.

Note that the normal operation of power transformers should be below the saturation region. However, when the transformer is operated beyond its rated power (during peak demand hours) or above nominal voltage (especially if power factor capacitor banks are left connected to the line under light load conditions), transformers are prone to operate under saturation.

Practically speaking, all transformers reach the saturation region on energization, developing large inrush (magnetizing) currents. Nevertheless, this is a condition that lasts only a few cycles. Another situation in which the power transformer may operate on the saturation region is under highly unbalanced load conditions; one of the phases carries a different current than the other phases, or the three phases carry unlike currents.

1.4 Fourier Series

By definition, a periodic function, *f(t)*, is that where $f(t) = f(t + T)$. This function can be represented by a trigonometric series of elements consisting of a DC component and other elements with frequencies comprising the fundamental component and its integer multiple frequencies. This applies if the following so-called Dirichlet conditions[2] are met:

If a discontinuous function, $f(t)$, has a finite number of discontinuities over the period T

If $f(t)$ has a finite mean value over the period T

If $f(t)$ has a finite number of positive and negative maximum values

The expression for the trigonometric series $f(t)$ is as follows:

$$f(t) = \frac{a_0}{2} \sum_{h=1}^{\infty} [a_h \cos(h\omega_0 t) + b_h \sin(h\omega_0 t)] \tag{1.8}$$

where $\omega_0 = 2\pi/T$.

We can further simplify Equation (1.8), which yields

$$f(t) = c_0 + \sum_{h=1}^{\infty} c_h \sin(h\omega_0 t + \phi_h) \tag{1.9}$$

where

$$c_0 = \frac{a_0}{2}, \quad c_h = \sqrt{a_h^2 + b_h^2}, \quad \text{and} \quad \phi_h = \tan^{-1}\left(\frac{a_h}{b_h}\right)$$

Equation (1.9) is known as a Fourier series and it describes a periodic function made up of the contribution of sinusoidal functions of different frequencies.

$(h\ \omega_0)$: hth order harmonic of the periodic function.

c_0: Magnitude of the DC component.

c_h and ϕ_h: Magnitude and phase angle of the hth harmonic component.

The component with $h = 1$ is called the fundamental component. Magnitude and phase angle of each harmonic determine the resultant waveform $f(t)$.

Equation (1.8) can be represented in a complex form as

$$f(t) = \sum_{h=1}^{\infty} c_h e^{jh\omega_0 t} \tag{1.10}$$

where $h = 0, \pm1, \pm2, \dots$.

$$c_h = \frac{1}{T} \int_{-T/2}^{T/2} f(t) e^{-jh\omega_0 t} dt \tag{1.11}$$

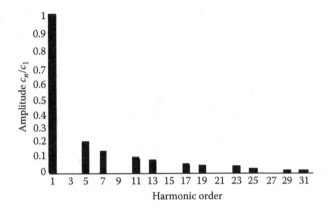

FIGURE 1.9
Example of a harmonic spectrum.

Generally, the frequencies of interest for harmonic analysis include up to the 40th or so harmonics.[3]

The main source of harmonics in power systems is the static power converter. Under ideal operation conditions, harmonics generated by a p pulse power converter are characterized by

$$Ih = \frac{I_1}{h}, \quad \text{and} \quad \cdots h = pn \pm 1 \tag{1.12}$$

where h stands for the characteristic harmonics of the load, $n = 1, 2, \ldots$, and p is an integer multiple of 6.

A bar plot of the amplitudes of harmonics generated in a six-pulse converter normalized as c_n/c_1 is called the harmonic spectrum, and it is shown in Figure 1.9.

The breakdown of the current waveform, including the four dominant harmonics, is shown in Figure 1.10. Notice that the harmonic spectrum is calculated with the convenient Electrotek Concepts TOP Output Processor.[4]

Noncharacteristic harmonics appear when:

The input voltages and currents are unbalanced.

The commutation reactance between phases is not equal.

The "space" between triggering pulses at the converter rectifier is not equal.

These harmonics are added together with the characteristic components and can produce waveforms with components that are not integer multiples of the fundamental frequency in the power system, also known as interharmonics.

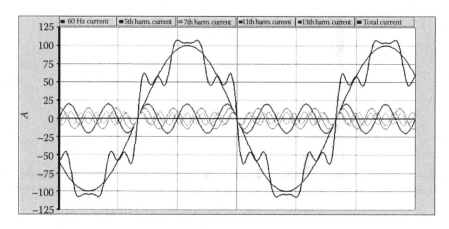

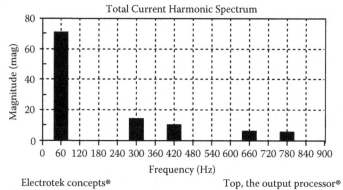

FIGURE 1.10
Decomposition of a distorted waveform.

A main source of interharmonics is the AC-to-AC converter, also called cycloconverter. These devices operate with a fixed amplitude and frequency at the input; at the output, amplitude and frequency can be variable. A typical application of a cycloconverter is as an AC traction motor speed control and other high-power, low-frequency applications, generally in the MW range.

Other important types of harmonics are those produced by electric furnaces, usually of a frequency lower than that of the AC system. These are known as subharmonics and are responsible for the light flickering phenomenon visually perceptible in incandescent and arc-type lighting devices.

Odd multiples of three (triplen) harmonics in balanced systems can be blocked using ungrounded neutral or delta-connected transformers because these are zero sequence harmonics. This is why triplen harmonics are often ignored in harmonic studies.

1.4.1 Orthogonal Functions

A set of functions, ϕ_i, defined in $a \leq x \leq b$ is called orthogonal (or unitary, if complex) if it satisfies the following condition:

$$\int_a^b \phi_i(x)\phi_j{}^*(x)dx = K_i\delta_{ij} \tag{1.13}$$

where $\delta_{ij} = 1$ for $i = j$, and 0 for $i \neq j$, and * is the complex conjugate.
It can also be shown that the functions:

$$\{1, \cos(\omega_0\ t), \ldots, \sin(\omega_0\ t), \ldots, \cos(h\omega_0\ t), \ldots, \sin(h\omega_0\ t), \ldots\} \tag{1.14}$$

for which the following conditions are valid,

$$\int_{-T/2}^{T/2} \cos kx \cos lx\ dx = \begin{cases} 0, k \neq l, \\ \pi, k = l, \end{cases} \tag{1.15}$$

$$\int_{-T/2}^{T/2} \sin kx \sin lx\ dx = \begin{cases} 0, k \neq l, \\ \pi, k = l, \end{cases} \tag{1.16}$$

$$\int_{-T/2}^{T/2} \cos kx \sin lx\ dx = 0 \cdots (k = 1,2,3,\ldots), \tag{1.17}$$

$$\int_{-T/2}^{T/2} \cos kx\ dx = 0 \cdots (k = 1,2,3,\ldots), \tag{1.18}$$

$$\int_{-T/2}^{T/2} \sin kx\ dx = 0 \cdots (k = 1,2,3,\ldots), \tag{1.19}$$

$$\int_{-T/2}^{T/2} 1dx = 2\pi \tag{1.20}$$

are a set of orthogonal functions. From Equation (1.14) to Equation (1.20), it is clear that the integral over the period ($-\pi$ to π) of the product of any two sine and cosine functions is zero.

1.4.2 Fourier Coefficients

Integrating Equation (1.8) and applying the orthogonal functions (Equations (1.15) to (1.20)), we obtain the Fourier coefficients as follows:

$$a_0 = \frac{2}{T} \int_{-T/2}^{T/2} f(t)dt, \tag{1.21}$$

$$a_h = \frac{2}{T} \int_{-T/2}^{T/2} f(t) \cos(h\omega_0 t)dt, \text{ and,} \tag{1.22}$$

$$b_h = \frac{2}{T} \int_{-T/2}^{T/2} f(t) \sin(h\omega_0 t)dt \tag{1.23}$$

where $h = 1, 2, \ldots, \infty$.

1.4.3 Even Functions

A function, $f(t)$, is called an even function if it has the following property:

$$f(t) = f(-t) \tag{1.24}$$

Figure 1.11 shows some examples of even functions.

1.4.4 Odd Functions

A function is called an odd function if

$$f(-t) = -f(t) \tag{1.25}$$

as portrayed in Figure 1.12.

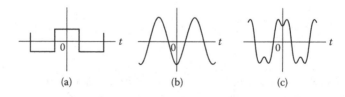

(a)   (b)   (c)

FIGURE 1.11
Examples of even functions.

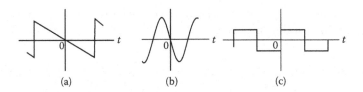

FIGURE 1.12
Examples of odd functions.

An even function is symmetrical to the vertical axis at the origin, and an odd function is asymmetrical to the vertical axis at the origin. A function with a period, T, is half-wave symmetrical if it satisfies the condition

$$f(t) = -f[t \pm (T/2)] \tag{1.26}$$

1.4.5 Effect of Waveform Symmetry

The Fourier series of an even function contains only cosine terms and may also include a DC component. Thus, the coefficients b_i are zero.

The Fourier series of an odd function contains only sine terms. The coefficients a_i are all zero.

The Fourier series of a function with half-wave symmetry contains only odd harmonic terms with $a_i = 0$ for $i = 0$ and all other even terms and $b_i = 0$ for all even values of i.

1.4.6 Examples of Calculation of Harmonics Using Fourier Series

1.4.6.1 Example 1

Consider the periodic function of Figure 1.13, which can be expressed as follows:

$$0, -T/2 < t < -T/4 \tag{1.27}$$

$$f(t) = 4, -T/4 < t < T/4 \tag{1.28}$$

$$0, T/4 < t < T/2 \tag{1.29}$$

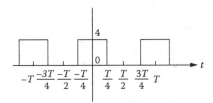

FIGURE 1.13
Square wave function.

for which we can calculate the Fourier coefficients using Equations (1.21) through (1.23) as follows:

$$a_0 = \frac{2}{T}\left(\int_{-T/2}^{T/2} f(t)dt\right) = \frac{2}{T}\left(\int_{-T/2}^{-T/4} 0 \cdot dt + \int_{-T/4}^{T/4} 4 \cdot dt + \int_{T/4}^{T/2} 0 \cdot dt\right)$$

(1.30)

$$= \frac{2}{T}[4(T/4 + T/4)] = 4$$

$$a_1 = \frac{2}{T}\int_{T-T/2}^{T/2} f(t) \cdot \cos(w_0 t)dt$$

$$= \frac{2}{T}\left(\int_{-T/2}^{-T/4} 0 \cdot \cos(\omega_0 t)dt + \int_{-T/4}^{+T/4} 4 \cdot \cos(\omega_0 t) + \int_{T/4}^{T/2} 0 \cdot \cos(\omega_0 t)dt\right) = \frac{8}{\pi}$$

(1.31)

We equally find that

$$a_i = \begin{cases} 0 - - - - - - - - - - - - - i = \text{even} \\ \qquad\qquad\qquad\qquad \cdots\cdots b_i = 0 \\ -1^{(i-1)/2}\frac{8}{i\pi} - - - - - - \quad i = \text{odd} \end{cases}$$

(1.32)

Therefore, from Equation (1.8), the Fourier series of this waveform is as follows:

$$f(t) = 2 + \frac{8}{\pi}\left(\cos\pi t - \frac{1}{3}\cos 3\pi t + \frac{1}{5}\cos 5\pi t - \ldots\ldots\right)$$

(1.33)

1.4.6.2 Example 2

Consider now that the periodic function of Figure 1.13 has its origin shifted one-fourth of a cycle, as illustrated in Figure 1.14.

$$a_0 = \frac{2}{T}\left(\int_{-T/2}^{T/2} f(t)dt\right) = \frac{2}{T}\left(\int_{-T/2}^{0} 0 \cdot dt + \int_{0}^{T/2} 4 \cdot dt\right)$$

(1.34)

$$= \frac{2}{T}[4(T/2 - 0)] = 4$$

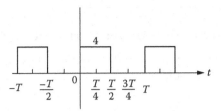

FIGURE 1.14
Square wave function shifted one fourth of a cycle relative to Figure 1.13.

$$a_1 = \frac{2}{T}\int_{-T/2}^{T/2} f(t)\cos(w_0 t)dt = \frac{2}{T}\left(\int_{-T/2}^{-0} 0 \cdot \cos(\omega_0 t)dt + \int_{0}^{T/2} 4 \cdot \cos(\omega_0 t)dt \right)$$

$$= \frac{8}{\omega_0 T}(\sin\omega_0 t)\Big|_0^{T/2} = \frac{8}{2\pi}\left[\sin\left(\frac{2\pi}{2}\right) - \sin(0) \right] = 0$$

(1.35)

Applying the orthogonality relations to Equation (1.22), we find that all a_i coefficients are zero. If we now try Equation (1.23), we determine the coefficients associated with the sine function in this series. For example, the first term, b_1, is calculated as follows:

$$b_1 = \frac{2}{T}\int_0^T f(t)\sin(\omega_0)t dt = \frac{2\omega_0}{2\pi}\int_0^{T/2} 4\sin(\omega_0 t)dt = -\left(\frac{2}{T}\right)\frac{4\cos(\omega_0 t)}{\omega_0}\Big|_0^{T/2} = \frac{8}{\pi}$$

(1.36)

Likewise, we find that:

$$b_i = \begin{cases} 0 ------ i = \text{even} \\ \dfrac{4}{i\pi} ------ i = \text{odd} \end{cases}$$

(1.37)

Therefore, following Equation (1.8), the Fourier series of this waveform reduced to its first three terms is as follows:

$$f(t) = 2 + \frac{8}{\pi}\left(\sin\pi t - \frac{1}{3}\sin 3\pi t + \frac{1}{5}\sin 5\pi t - .... \right)$$

(1.38)

1.5 Power Quality Indices under Harmonic Distortion

This section describes power quality indices that are comprehensibly described in references 5 and 6.

1.5.1 Total Harmonic Distortion

Total harmonic distortion (THD) is an important index widely used to describe power quality issues in transmission and distribution systems. It considers the contribution of every individual harmonic component on the signal. THD is defined for voltage and current signals, respectively, as follows:

$$THD_V = \frac{\sqrt{\sum_{h=2}^{\infty} V_h^2}}{V_1} \tag{1.39}$$

$$THD_I = \frac{\sqrt{\sum_{h=2}^{\infty} I_h^2}}{I_1} \tag{1.40}$$

This means that the ratio between rms values of signals, including harmonics and signals considering only the fundamental frequency, defines the total harmonic distortion.

1.5.2 Total Demand Distortion

Harmonic distortion is most meaningful when monitored at the point of common coupling (PCC)—usually the customer's metering point—over a period that can reflect maximum customer demand, typically 15 to 30 min, as suggested in IEEE 519.[7] Weak sources with a large demand current relative to their rated current will tend to show greater waveform distortion. Conversely, stiff sources characterized for operating at low demand currents will show decreased waveform distortion. The total demand distortion is based on the demand current, I_L, over the monitoring period:

$$TDD = \frac{\sqrt{\sum_{h=2}^{\infty} I_h^2}}{I_L} \tag{1.41}$$

1.5.3 Telephone Influence Factor TIF

This index is found in IEEE 519[7] as a measure of audio circuit interference produced by harmonics in electric power systems. It will thus use the total harmonic distortion concept influenced by appropriate weighting factors, ω_h, that establish the sensitivity of the human ear to noise from different frequencies:

$$\text{TIF} = \frac{\sqrt{\sum_{h=2}^{\infty} w_h^2 I_h^2}}{I_{rms}} \tag{1.42}$$

1.5.4 C Message Index

This index is related in reference 7 to communication interference:

$$C_I = \frac{\sqrt{\sum_{h=2}^{\infty} c_h^{\,2} I_h}}{I_{rms}} \tag{1.43}$$

where c_h is the weighting factor, ω_h, divided by five times the harmonic order h.

1.5.5 *I* * *T* and *V* * *T* Products

These indices are used as another measure of harmonic interference in audio circuits. Because of their intimate relation with total waveform distortion, $I * T$ and $V * T$ are also indicative of shunt capacitor stress and voltage distortion, respectively:

$$I * T = \text{TIF} * I_{rms} = \sqrt{\sum_{h=2}^{\infty} (w_h I_h)^2} \tag{1.44}$$

$$V * T = \text{TIF} * V_{rms} = \sqrt{\sum_{h=2}^{\infty} (w_h V_h)^2} \tag{1.45}$$

1.5.6 K Factor

This is a useful index intended to follow the requirements of the National Electrical Code (NEC) and Underwriters Laboratories (UL) (well summarized by its originator, Frank[8]) regarding the capability of distribution and special application transformers in industry to operate within specified thermal limits in harmonic environments. These are transformers designed to operate at lower flux densities than conventional designs to allow for the additional flux produced by (largely the third) harmonic currents. Also, to reduce the eddy or circulating current losses in the core, strip windings, interleaving windings, and transposition conductors are used. The formula used to calculate the K factor (as presented in the *IEEE Tutorial on Modeling and Simulations*[5]) is as follows:

$$K = \frac{\sum_{h=1}^{\infty} h^2 \left(\frac{I_h}{I_1}\right)^2}{\sum_{h=1}^{\infty} \left(\frac{I_h}{I_1}\right)^2} = \sum_{h=1}^{\infty} [I_h(p.u.)]^2 (h^2) \tag{1.46}$$

1.5.7 Displacement, Distortion, and Total Power Factor

With an increasing harmonic distortion environment, the conventional definition of power factor as the cosine of the angle between fundamental frequency voltage and current has progressed to consider the signal's rms values, which make up the contribution of components of different frequencies. Thus, displacement power factor (DPF) continues to characterize the power frequency factor, while distortion (or true) power factor (TPF) emerges as the index that tracks rms signal variations. Total power factor (PFtotal) thus becomes the product of distortion and true power factors:

$$PF_{\text{total}} = DPF * TPF = \cos(\theta 1 - \delta 1) * \frac{P_{\text{total}}}{S_{\text{total}}}$$

$$= \left(\frac{P_1}{V_1 I_1}\right) \left(\frac{\sum_{h=1}^{\infty} V_h I_h \cos(\theta_h - \delta_h)}{\sqrt{\sum_{h=1}^{\infty} (V_h)^2} \sqrt{\sum_{h=1}^{\infty} (I_h)^2}}\right) = \left(\frac{P_1}{V_1 I_1}\right) \sqrt{\frac{1}{1 + \left(\frac{\text{THD}_I}{100}\right)^2}} \tag{1.47}$$

where P_1, V_1, and I_1 are fundamental frequency quantities and V_h, I_h, θ_h, and δ_h are related to a frequency, h, times the system power frequency.

Because true power factor is always less than unity, it also holds that

$$PF_{\text{total}} \leq DPF \tag{1.48}$$

In Equation (1.47), note that fundamental displacement power factor is the ratio between $P_{\text{total}}/S_{\text{total}}$ or $P_1/(V_1 I_1)$

1.5.8 Voltage-Related Parameters

Crest factor, unbalance factor, and flicker factor are intended for assessing dielectric stress, three-phase circuit balance, and source stiffness with regard to its capability of maintaining an adequate voltage regulation, respectively:

$$\text{Crest Factor} = \frac{V\text{peak}}{V\text{rms}} \tag{1.49}$$

$$\text{Unbalance Factor} = \frac{|V_-|}{|V_+|} \tag{1.50}$$

$$\text{Flicker Factor} = \frac{\Delta V}{|V|} \tag{1.51}$$

1.6 Power Quantities under Nonsinusoidal Situations

This section describes the Institute of Electrical and Electronics Engineers (IEEE) quantities under nonsinusoidal situations.[5] It is noteworthy to emphasize that all quantities referred to in this section are based on the trigonometric Fourier series definition described in Section 1.4 as Equation (1.9). These quantities are expressed in a way that they account for the contribution of individual harmonic frequency components. In this section, $f(t)$ represents instantaneous voltage or current as a function of time; F_h is the peak value of the signal component of harmonic frequency h.

1.6.1 Instantaneous Voltage and Current

$$f(t) = \sum_{h=1}^{\infty} f_h(t) = \sum_{h=1}^{\infty} \sqrt{2} F_h \sin(h\omega_0 t + \theta_h) \tag{1.52}$$

1.6.2 Instantaneous Power

$$p(t) = v(t)i(t) \tag{1.53}$$

1.6.3 rms Values

$$F_{rms} = \sqrt{\frac{1}{T}\int_0^T f^2(t)dt} = \sqrt{\sum_{h=1}^{\infty} F_h^2} \tag{1.54}$$

where F_{rms} is the root mean square of function F, which in our case can be voltage or current.

1.6.4 Active Power

Every harmonic provides a contribution to the average power that can be positive or negative. However, the resultant harmonic power is very small relative to the fundamental frequency active power.

$$P = \frac{1}{T}\int_0^T p(t)dt = \sum_{h=1}^{\infty} V_h I_h \cos(\theta_h - \delta_h) = \sum_{h=1}^{\infty} P_h \tag{1.55}$$

1.6.5 Reactive Power

$$Q = \frac{1}{T}\int_0^T q(t)dt = \sum_{h=1}^{\infty} V_h I_h \sin(\theta_h - \delta_h) = \sum_{h=1}^{\infty} Q_h \tag{1.56}$$

1.6.6 Apparent Power

Many arguments have been written about the most suitable way to express the apparent power under the presence of harmonic distortion. A good summary of such efforts can be found in Arrillaga and Watson,[6] who refer to the initial approach by Budeanu,[9] Fryze,[10] and Kusters and Moore[11] and the most recent work by Emanuel.[12,13] Arrillaga and Watson also show how all formulations lead to somewhat different results in the determination of reactive power quantities under waveform distortion. An expression generally accepted by IEEE and the International Electrotechnical Commission (IEC) is that proposed by Budeanu in Antoniu[9]:

$$S^2 = P^2 + \sum_{i=1}^{n} V_1 I_1 \sin(\varphi_1) + D^2 \tag{1.57}$$

For three-phase systems, the per-phase (k) vector apparent power, S_v, as proposed in Frank,[8] can be expressed, as adapted from Arrillaga and

Watson,[6] as follows:

$$S_v = \sqrt{\left(\left(\sum_k P_k\right)^2 + \left(\sum_k Q_{bk}\right)^2 + \left(\sum_k D_k\right)^2\right)} \qquad (1.58)$$

and the arithmetic apparent power, S_a, as

$$S_a = \sum_k \sqrt{P_k^2 + Q_{bk}^2 + D_k^2} \qquad (1.59)$$

where P, Q_b, and D are the active, reactive, and distortion orthogonal components of power, respectively.

From Antoniu,[9] the expression for the per-phase apparent rms power, S_e, as adapted in Arrillaga and Watson[6] is

$$S_e = \sum_k \sqrt{(P_k^2 + Qf_k^2)} = \sum_k V_k I_k \qquad (1.60)$$

and the apparent power for a three-phase system, S_s:

$$S_s = V_{rms} I_{rms} = \sqrt{P^2 + Q_f^2} = \sqrt{\sum_k V_k^2} \sqrt{\sum_k I_k^2} \qquad (1.61)$$

where Q_f is the reactive power.

Emanuel[12] is an advocate for the separation of power in fundamental and nonfundamental components and further proposes the determination of apparent power, S, as

$$S = \sqrt{S_1^2 + S_n^2} \qquad (1.62)$$

where S_1 is the fundamental and S_n the nth component of apparent power. The harmonic active power, P_H, embedded in S_n is negligible, around half a percent of the fundamental active power, according to Kusters and Moore.[11]

1.6.7 Voltage in Balanced Three-Phase Systems

Harmonics of different order form the following sequence set:

Positive sequence: 1, 4, 7, 10, 13, …

Negative sequence: 2, 5, 8, 11, 14, …

Zero sequence: 3, 6, 9, 12, 15, … (also called triplen)

The positive sequence system has phase order R, S, T (a, b, c), and the negative sequence system has phase order R, T, S (a, c, b). In the zero

TABLE 1.1

Phase Sequences of Harmonics in a Three-Phase Balanced System

Harmonic order	1	2	3	4	5	6	7	8	9	10	11 ...
Phase sequence	Positive	Negative	Zero	Positive	Negative	Zero	Positive	Negative	Zero	Positive	Negative ...

sequence system, the three phases have an equal phase angle. This results in a shift for the harmonics, which for a balanced system can be expressed as follows:

$$Va_h(t) = \sqrt{2}V_h\sin(h\omega_0 t + \theta_h) \tag{1.63}$$

$$Vb_h(t) = \sqrt{2}V_h\sin\left(h\omega_0 t - \frac{2h\pi}{3} + \theta_h\right) \tag{1.64}$$

$$Vc_h(t) = \sqrt{2}V_h\sin\left(h\omega_0 t + \frac{2h\pi}{3} + \theta_h\right) \tag{1.65}$$

Note that in Equation (1.60) through Equation (1.62), harmonics voltages and currents are displaced 120° from one another. The phase sequences of harmonic voltage or currents can be described as in Table 1.1. Notice that triplen harmonics are zero sequence.

In an unbalanced system, harmonic currents will contain phase sequences different from those in Table 1.1.

1.6.8 Voltage in Unbalanced Three-Phase Systems

Unbalanced voltage conditions are rare but possible to find in three-phase electric power systems. The main reason for voltage unbalance is an uneven distribution of single-phase loads; other reasons may include mutual effects in asymmetrical conductor configurations. During load or power system unbalance, it is possible to find voltages of any sequence component:

$$\begin{bmatrix} V1_h \\ V2_h \\ V3_h \end{bmatrix} = \frac{1}{3}\begin{bmatrix} 1 & a & a^2 \\ 1 & a^2 & a \\ 1 & 1 & 1 \end{bmatrix}\begin{bmatrix} Va_h \\ Vb_h \\ Vc_h \end{bmatrix} \tag{1.66}$$

where $a = e^{j120°}$.

In most cases, there is a dominant sequence component with a meager contribution from other frequencies. Under certain conditions involving triplen harmonics, there can be only positive or negative sequence components.

References

1. De Almeida, A., Understanding Power Quality, *Home Energy Magazine Online*, November/December 1993, http://homeenergy.org/archive/hem.dis.anl.gov/eehem/93/931113.html.
2. Edminster, J.A., *Electrical Circuits*, Schaum's Series, McGraw Hill, New York, 1969.
3. IEC 61000-4-7, *Electromagnetic Compatibility (EMC)—Part 4-7: Testing and Measurement Techniques—General Guide on Harmonics and Interharmonics Measurements and Instrumentation, for Power Supply Systems and Equipment Connected Thereto*, Edition 2, 2002.
4. Electrotek Concepts, TOP—The Output Processor, http://www.pqsoft.com/top/.
5. IEEE Power Engineering Society, *IEEE Tutorial on Modeling and Simulations*, IEEE PES, 1998.
6. Arrillaga, J., and Watson, N., *Power Systems Harmonics*, 2nd ed., Wiley, New York, 2003.
7. IEEE 519-1992, *Recommended Practices and Requirements for Harmonic Control in Electric Power Systems*.
8. Frank, J.M., Origin, Development and Design of K-Factor Transformers, *IEEE Ind. Appl. Mag.*, September/October 1997.
9. Antoniu, S., Le Régime Energique Deformant. Une Question de Priorité, *RGE*, 6/84, 357–362, 1984.
10. Fryze, S., Wirk, Blind und Scheinleistung in Electrischen Stromkreisien mit Nitchsinuformigen Verlauf von Strom und Spannung, *Electrotechnisch Zeitschrift*, 596–599, June 1932.
11. Kusters, N.L., and Moore, W.J.M., On Definition of Reactive Power under Nonsinusoidal Conditions, *IEEE Trans. Power Appar. Syst.*, 99, 1845–1850, 1980.
12. Emanuel, A.E., Power in Nonsinusiodal Situations, a Review of Definitions and Physical Meaning, *IEEE Trans. Power Delivery*, 5, 1377–1383, 1990.
13. Emanuel, A.E., Apparent Power Components and Physical Interpretation, in *International Conference on Harmonics and Quality of Power (ICHQP '98)*, Athens, 1998, 1–13.

2

Harmonic Sources

2.1 Introduction

Although power system harmonics is the topic of this book, it is important to stress that harmonic waveform distortion is just one of many different disturbances that perturb the operation of electrical systems. It is also a unique problem in light of an increasing use of power electronics that basically operate through electronic switching. Fortunately, the sources of harmonic currents seem to be sufficiently well identified, so industrial, commercial, and residential facilities are exposed to well-known patterns of waveform distortion.

Different nonlinear loads produce different but identifiable harmonic spectra. This makes the task of pinpointing possible culprits of harmonic distortion more tangible. Utilities and users of electric power have to become familiar with the signatures of different waveform distortions produced by specific harmonic sources. This will facilitate the establishment of better methods to confine and remove them at the sites where they are produced. In doing this, their penetration in the electrical system affecting adjacent installations will be reduced. As described in Chapters 6 and 8, parallel resonant peaks must be properly accounted for when assessing waveform distortion. Otherwise, a filtering action using single-tuned filters to eliminate a characteristic harmonic at a given site may amplify the waveform distortion if the parallel peak (pole) of the filter coincides with a lower-order characteristic harmonic of the load. Active filters may overcome this hurdle, but they must be well justified to offset their higher cost.

The assessment of harmonic propagation in a distribution network, on the other hand, requires an accurate representation of the utility source. Weak sources will be associated with significant harmonic distortion that can in turn affect large numbers of users served from the same feeder that provides power to the harmonic-producing customer. This will become particularly troublesome when harmonics are created at more than one location—for example, in a cluster of industrial facilities served from the same feeder.

Thus, utilities may be inadvertently degrading the quality of power by serving heavy harmonic producers from a weak feeder.

From the perspective of the customer, power quality means receiving a clean sinusoidal voltage waveform with rms variations and total harmonic distortion within thresholds dictated by a number of industrial standards. Often, however, utilities find it difficult to keep up with these regulations. The culprit is often found on the customer loads, which from victims they turn into offenders when they draw large blocks of currents from the AC source in slices. This occurs whenever they convert power from one form into another through rectification and inversion processes. The waveform chopping process leads to noise-like structures often regarded as dirty, unclean, or polluted power. This is nothing but the harmonic distortion of the voltage supply, which is the subject matter of this book and must be assessed from all possible perspectives.

As it will be described, the main effects of this distortion range from increased equipment losses that shorten the lifetime of equipment like transformers and cables to interference in audio and data communication to possible protective devices' nuisance tripping. The cost involved in cleaning this harmonic noise will often counteract the benefit obtained from improving equipment and appliances to better operate them in disturbed environments. As a general rule, the more sophisticated or sensitive the electronic equipment, the higher the cost to keep it running given its increased sensitivity to power quality disturbances. Sensitive industrial processes, such as highly automated assembly lines, are prone to power-related damage from severe harmonic distortion. On the other hand, the stiffer the AC source, the lower the voltage distortion across the source terminals given the reduced voltage drop in a diminished source impedance.

Harmonic distortion is no longer a phenomenon confined to industrial equipment and processes, where the first power quality concerns developed. Uninterruptible power supplies (UPSs), personal computers (PCs), and electronic and entertaining devices proliferate nowadays in commercial and residential installations. These special kinds of loads represent formidable sources of harmonic currents, and they increase with the expanding use of entertaining devices, including TV sets, music amplifiers, video recorders, battery charges for smart phones and tablets, digital clocks, data centers, and other sensitive electronic equipment.

The interaction between power utility AC sources and customer loads will be further affected by distributed resources (often known as distributed generators (DGs)) that employ electronic switching technologies, increasingly used by utilities and industry to better cope with peak demand. The expanding presence of this type of DGs, the integration of renewable wind and solar power, high-voltage direct current (HVDC), and smart devices will contribute to an overall rise in harmonic distortion in electric distribution networks. An increased effort by utilities and industries alike to devise improved mitigation methods that can keep harmonic distortion within allowable limits is foreseen.

2.2 The Signature of Harmonic Distortion

Figure 2.1 illustrates a simple PSCAD (Power Systems Computer-Aided Design)[1] model to produce distortion on the voltage waveform. The student edition of this software has been used for this and other examples presented throughout the book.[1] Consequently, only simple cases in which modeling can be achieved with the reduced number of nodes available are considered in the examples. For larger applications, the student edition falls short of dimensions to model all relevant features of an electrical installation and complex loads. Though simplified, the model in Figure 2.1 provides a practical glance at the effects of harmonic currents of different frequencies on AC voltage waveform signatures. The harmonic injected current was kept constant, and the simulation included harmonic currents of the 2nd to the 11th order.

Figure 2.2(a) to (j) shows the corresponding results. Even (not odd) harmonics are unlikely to exist in balanced systems; however, in this example, we include them to get an overall idea of the different waveform distortion signatures. Even harmonics may arise (for instance, under waveform asymmetry) if thyristor triggering angles are slightly different on every half cycle. This is by no means a comprehensive assortment of all the harmonic distortion signatures likely to be found in practical situations. Just by varying the phase angle and amplitude of the harmonic currents relative to the fundamental frequency, we would assemble an endless collection of different distorted waveforms.

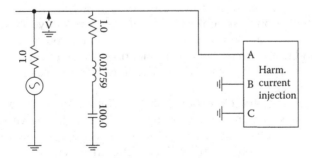

FIGURE 2.1
Harmonic generator to determine the voltage waveform distortion at different harmonic frequencies.

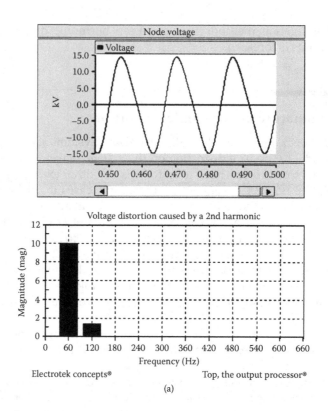

(a)

FIGURE 2.2
Waveform distortion imposed by currents of harmonic orders 2 through 11. *(Continued)*

2.3 Traditional Harmonic Sources

Prior to the development of power electronic switching devices, harmonic current propagation was looked at from the perspective of design and operation of power apparatus devices with magnetic iron cores, like electric machines and transformers. In fact, at that time the main source of harmonics must have involved substation and customer transformers operating in the saturation region.

Nowadays, harmonic distortion produced under transformer saturation probably at peak demand or under elevated voltage during very light load conditions is only one of numerous situations that generate harmonic waveform distortion. Possibly, electric furnaces should be regarded as the second most important cause of concern in high-power applications in industry, second to power converter utilization.

The sources of waveform distortion in power systems are multiple, and in industrial installations, they can be found from small (less than 1 kVA)

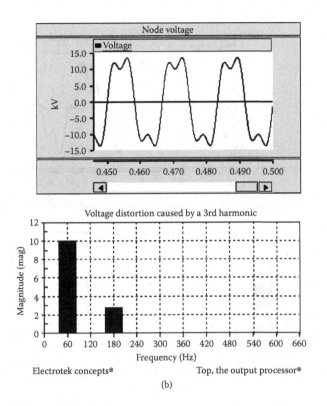

(b)

FIGURE 2.2 (Continued)
Waveform distortion imposed by currents of harmonic orders 2 through 11. *(Continued)*

to several tens of megavoltamperes. However, as mentioned earlier, commercial and residential facilities can also become significant sources of harmonics. This is particularly true when the combined effects of all individual loads served by the same feeder are taken into account. For instance, a simple power source of a home desktop computer may draw around 4 A from a 127 V main, or around 500 VA. A medium-voltage feeder typically serving around 2500 low voltage (LV) customers would be eventually supplying around 1.25 MVA of computer power under the likely scenario of having all customers checking e-mail accounts or browsing on the Internet in unison after dinnertime, for example.

A six-pulse converter shows a theoretical amplitude (as shortly addressed in this chapter) of around 20% just for the fifth harmonic current, as it was depicted in Figure 1.1. This translates into a similar percentage of reactive power. In our example, this amounts to around 1.25 × 0.20 or 250 kVA of 300 Hz power that can be envisaged as current injection back into the power system. This contributes to the distortion of the voltage waveform. Other harmonics generated by the power converter would add additional distortion to

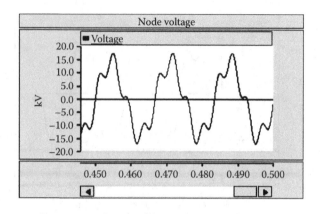

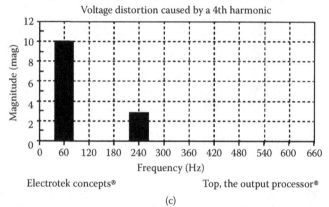

Voltage distortion caused by a 4th harmonic

Electrotek concepts®   Top, the output processor®

(c)

FIGURE 2.2 (Continued)
Waveform distortion imposed by currents of harmonic orders 2 through 11. *(Continued)*

the voltage waveform because they can be conceived as individual spectral currents injected into the power system.

As previously mentioned, some harmonic sources like saturated transformers have existed from the early times when electricity was first transformed and distributed over power lines for commercial purposes. As described in the next section, the operation of transformers near the saturation zone is the result of excessive magnetic flux through the core, which limits the linear increase of the magnetic flux density. Rotating machines are another example of equipment that may behave as a harmonic distortion source under overloading conditions.

The use of electricity involving loads that require some form of power conditioning like rectification or inversion is on the rise, as mentioned in Chapter 1. The greatest majority of industrial nonlinear loads are related to solid-state switching devices used in power converters that change electric power from one form to another. This includes, among others, AC-to-DC energy conversion for DC motor speed control, and AC to DC and back to

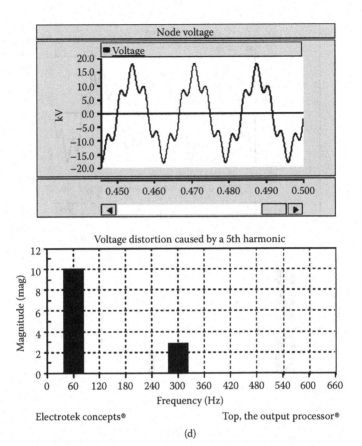

FIGURE 2.2 (Continued)
Waveform distortion imposed by currents of harmonic orders 2 through 11. *(Continued)*

AC at variable frequencies for processes involving speed control of induction motors. Most bulk energy conversion processes take place in the oil,
mining, steel mill, pulp and paper, textile, and automobile industries. Other
applications include manufacturing assembly lines and electrolytic coat-
ing processes, which can produce significant amounts of harmonic current
generation.

Arc devices (namely, electric furnaces, soldering equipment, fluorescent,
and mercury vapor or high-pressure sodium lamps) can become very special
sources of harmonics in that they can involve frequencies below the fundamental power frequency and fractional harmonics. The former are called
subharmonics, and the latter are known as interharmonics. Subharmonic
generation can typically take place when arc-type devices are sourced
through weak transmission or distribution systems, i.e., those with small
ratios of short circuit to load current.

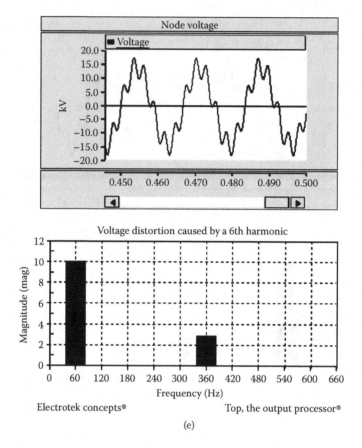

FIGURE 2.2 (Continued)
Waveform distortion imposed by currents of harmonic orders 2 through 11. *(Continued)*

Large inrush currents during switching of capacitor banks, transformers, and rotating machines into the distribution system can develop harmonic currents. IEEE 519[2] allows for harmonic distortion limits 50% higher than recommended values during start-ups and unusual conditions lasting less than 1 h. Harmonic distortion due to inrush currents on transformer energization and to outrush currents developed under shunt connection of capacitor banks (especially when more capacitors are added to an existing bank) fall in this category. In the latter case, large currents are discharged from one bank into the other because the only limiting element existing between the two banks is the surge impedance of the connecting cable. This may explain some nuisance operations of capacitor bank fuses.

Other harmonic sources may include ferroresonance phenomena, which may remain undetected for periods of minutes and even hours until reported by affected customers. Thus, the severe waveform distortion developed no longer fits into the IEEE 519 definition for unusual conditions. This undesired

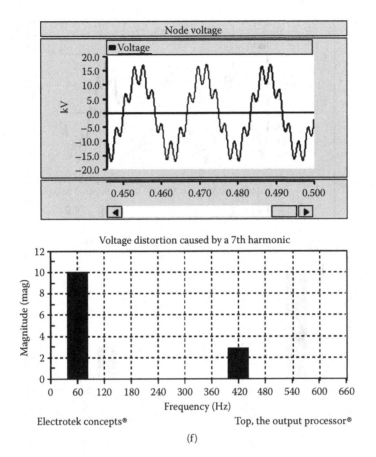

FIGURE 2.2 (Continued)
Waveform distortion imposed by currents of harmonic orders 2 through 11. *(Continued)*

event has the potential to produce extensive damage to customer facilities and equipment, and therefore utilities and industry must minimize the risk to trigger this condition. American[3] and European[4] publications address this phenomenon and provide guidance to assess the possibility of ferroresonance involving small transformers fed off from underground cables.

2.3.1 Transformers

A transformer can incur in core saturation conditions in either of the following cases:

When operating above rated power
When operating above rated voltage

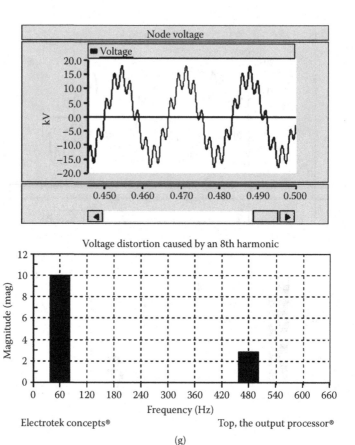

FIGURE 2.2 (Continued)
Waveform distortion imposed by currents of harmonic orders 2 through 11. *(Continued)*

The first situation can arise during peak demand periods, and the second case can occur during light load conditions, especially if utility capacitor banks are not disconnected accordingly and the feeder voltage rises above nominal values.

A transformer operating on the saturation region will show a nonlinear magnetizing current similar to that illustrated in Figure 2.3, which contains a variety of odd harmonics, with the third dominant. The effect will become more evident with increasing loading. In an ideal lossless core, no hysteresis losses are produced. The magnetic flux and the current needed to produce them are related through the magnetizing current of the steel sheet material used in the core construction. Even under this condition, if we plot the magnetizing current vs. time for every flux value considered, the resultant current waveform would be far from sinusoidal.

When the hysteresis effect is considered, this nonsinusoidal magnetizing current is not symmetrical with respect to its maximum value. The distortion

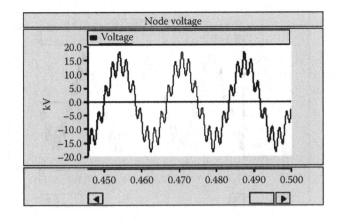

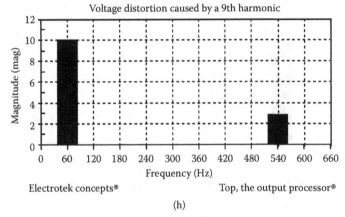

(h)

FIGURE 2.2 (Continued)
Waveform distortion imposed by currents of harmonic orders 2 through 11. *(Continued)*

is typically due to triplen harmonics (odd multiples of 3, namely, the 3rd, 9th, 15th, etc.), but mainly due to the third harmonic. This spectral component can be confined within the transformer using delta transformer connections. This will help maintain a supply voltage with a reasonable sinusoidal waveform.

In three-legged transformers, the magnetomotive forces (mmfs) of triplen harmonics are all in phase and act on every leg in the same direction. Therefore, the trajectory of the magnetic flux for the triplen harmonics extends outside the boundaries of the core. The high reluctance of this trajectory reduces the flux of triplen harmonics to a very small value. The components of fifth and seventh harmonics can be considerable (5 to 10%) to produce considerable distortion and ought not to be ignored.

In electric power distribution networks, harmonics due to transformer magnetizing current reach their maximum value early before dawn when the system is lightly loaded and voltage level is high. When a transformer is deenergized, it is possible that it retains residual magnetic flux in the core.

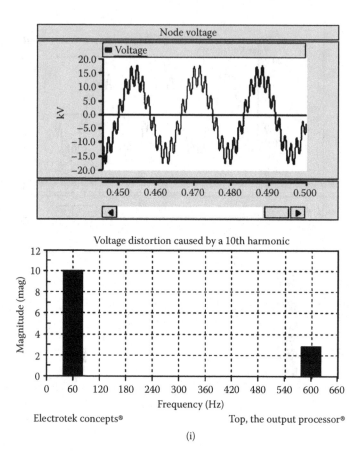

(i)

FIGURE 2.2 (Continued)
Waveform distortion imposed by currents of harmonic orders 2 through 11. *(Continued)*

On reenergization, this flux coalesces with the magnetizing flux produced by the inrush current, and the two combined can yield peak values three times or higher the nominal flux at rated load. The resulting effect may cause the transformer core to reach extreme saturation levels involving excessive ampere turns within the core. Consequently, magnetizing currents as large as 5 to 10 per unit (p.u.) of nominal current (compared with 1 to 2% of nominal magnetizing current during steady-state operating conditions) can develop. The duration of the magnetizing current is mainly a function of the primary winding resistance. For large transformers with large winding resistance, this current can remain for many seconds.

As it has been mentioned above, the harmonic content of steady-state currents in three-phase systems does not involve even harmonics. These can appear under waveform asymmetry when the positive and negative half cycles are not of the same amplitude. However, under energization, a

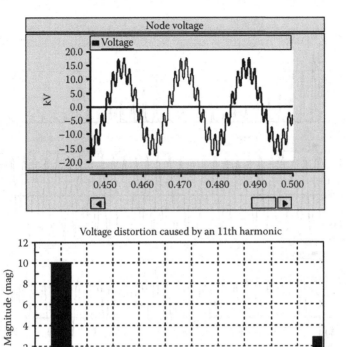

Voltage distortion caused by an 11th harmonic

Electrotek concepts®   Top, the output processor®

(j)

FIGURE 2.2 (Continued)
Waveform distortion imposed by currents of harmonic orders 2 through 11.

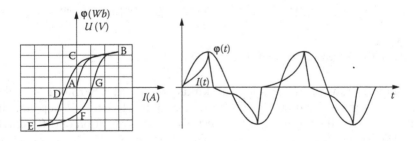

FIGURE 2.3
Distorted current under transformer saturation conditions.

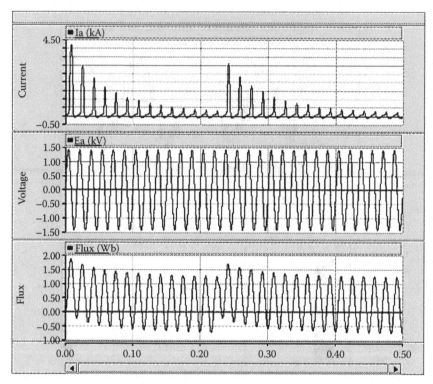

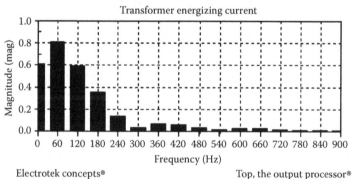

FIGURE 2.4
Typical transformer energizing current.

distribution transformer develops all kinds of low-order harmonics involving even harmonics (notably the second and the fourth, as depicted in Figure 2.4), which are often used for restraining the operation of differential protection.

The transformer model used in the PSCAD software is the so-called unified magnetic equivalent circuit (UMEC) transformer,[5,6] which overcomes

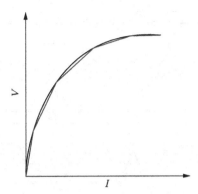

FIGURE 2.5
Piecewise linear representation of transformer conductance.

the problem of pulling together the information on the transformer core and winding characteristics by deriving the elements of the inductance matrix from test data on open- and short-circuit tests. UMEC simulates the nonlinear characteristic of the core by using a piecewise linear representation of the equivalent branch conductance, as shown in Figure 2.5. This allows reducing the processing time by shortening the number of matrix inversions.

The presence of the even harmonics and their decaying nature are typical under transformer saturation. Here we can reproduce the harmonics created during transformer saturation by injecting harmonic currents similar to the typical harmonic spectrum of Figure 2.4 into an AC source and a short feeder representation. Figure 2.6 depicts the simplified model re-created in PSCAD. The load voltage and current waveforms and their harmonic spectra are shown in Figure 2.7 for a 180° phase angle between fundamental and harmonic currents.

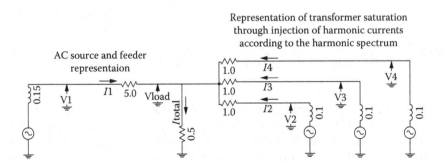

FIGURE 2.6
Simplified way to simulate harmonic injection into an AC source, typical of transformer saturation.

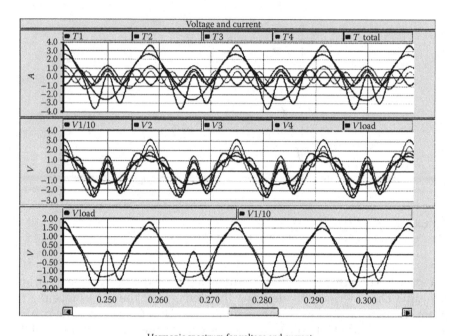

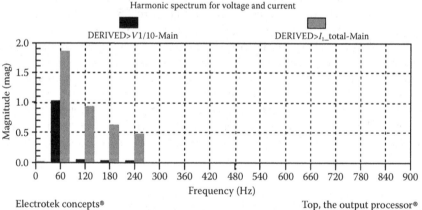

FIGURE 2.7
Waveform distortion due to harmonic currents, typical of transformer saturation.

Resultant total harmonic distortion (THD) values for source voltage (V_1) and load current (I_{total}) are 6 and 66%, respectively. Note on the bottom plot the distorted voltage waveform of the load relative to the voltage waveform of the source, which also undergoes some distortion. Figure 2.8 shows corresponding results for a weak AC source with impedance around 50% larger than that considered in generating the results in Figure 2.7. In the latter case, THD values resulted in 9 and 98% for voltage and current, respectively, *a*

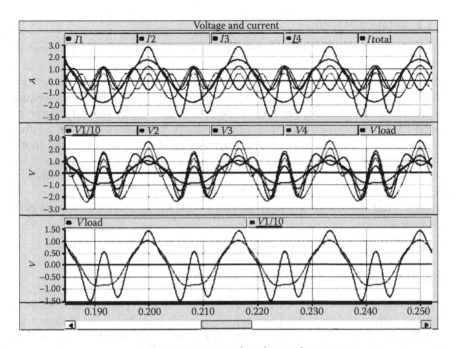

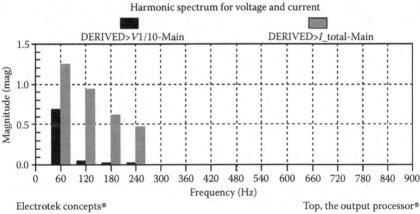

FIGURE 2.8
Waveform distortion due to harmonic currents, typical of transformer saturation in a weaker system.

substantial increase. This illustrates the important role that the source impedance plays in determining the voltage distortion levels.

As noticed, both voltage and current waveforms suffer from distortion produced by harmonic currents during transformer saturation phenomena. Transformer saturation can also take place following a voltage dip because a sudden change in voltage leads to a DC component in the magnetizing flux.

2.3.2 Rotating Machines

As a result of small asymmetries on the machine stator or rotor slots or slight irregularities in the winding patterns of a three-phase winding of a rotating machine, harmonic currents can develop. These harmonics induce an electromotive force (emf) on the stator windings at a frequency equal to the ratio of speed/wavelength. The resultant distribution of magneto-motive forces (mmfs) in the machine produces harmonics that are a function of speed. Additional harmonic currents can be created upon magnetic core saturation. However, these harmonic currents are usually smaller than those developed when the machines are fed through variable frequency drives (VFDs).

Additional discussion on harmonics in rotating machines is provided in Chapter 4.

2.3.3 Power Converters

The increasing use of the power conditioners in which parameters like voltage and frequency are varied to adapt to specific industrial and commercial processes has made power converters the most widespread source of harmonics in distribution systems. Electronic switching helps the task to rectify 50/60 Hz AC into DC power. In DC applications, the voltage is varied through adjusting the firing angle of the electronic switching device. Basically, in the rectifying process, current is allowed to pass through semiconductor devices during only a fraction of the fundamental frequency cycle, for which power converters are often regarded as energy-saving devices. If energy is to be used as AC but at a different frequency, the DC output from the converter is passed through an electronic switching inverter that brings the DC power back to AC, but of the desired frequency.

Converters can be grouped into the following categories:

Large power converters like those used in the metal smelter industry and in HVDC transmission systems.

Medium-size power converters like those used in the manufacturing industry for motor speed control and in the railway industry.

Small power rectifiers used in residential entertaining devices, including TV sets and personal computers. Battery chargers are another example of small power converters.

Figure 2.9(a) describes the basic relation between current and voltage in which a halfway control uses a gated turn-off thyristor (GTO), as portrayed by Finney[7], to draw current during part of the AC waveform positive half cycle. A strong DC component due to the switching action taking place only on one side of the AC cycle is manifest in Figure 2.9(b). In addition,

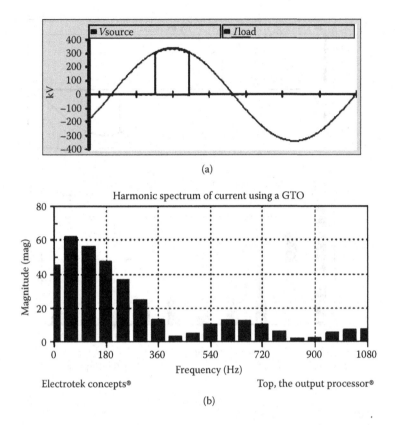

(a)

(b)

Electrotek concepts®

Top, the output processor®

FIGURE 2.9
AC switch using a GTO and harmonic spectrum of the current.

zero-sequence (triplen) and even harmonic components are evident. This peculiar behavior is also characteristic of unbalanced three-phase systems.

If a different switching device like an insulated gate bipolar transistor (IGBT), as described by Finney,[7] in which current flows during only part of the time on every half cycle (Figure 2.10(a)), is used, the current harmonic spectrum of Figure 2.10(b) is obtained.

Notice how, in this case, the waveform does not contain even harmonics due to the symmetry of the switching action relative to the x-axis. Therefore, only odd and zero-sequence harmonics show up. The same result can be achieved through Fourier analysis decomposition of any waveform containing identical features on the two half cycles.

To further illustrate the power converter as a harmonic source, let us refer to the six-pulse VFD of Figure 2.11. Harmonic currents, i_{hL}, produced by the rotating machine will be practically confined to the load side, beyond the DC bus. This is regardless of whether the converter is of a current source (a) or a voltage source (b) configuration. Due to the commutation of current

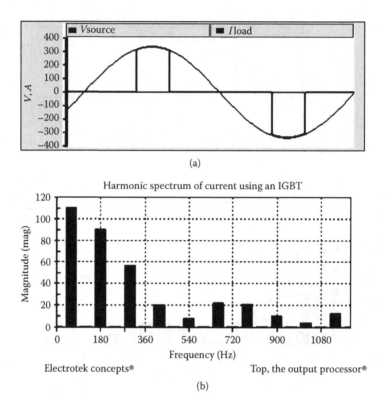

(a)

(b)

FIGURE 2.10
AC switch using an IGBT and harmonic spectrum of the current.

from one phase to another during the rectifying process on the converter, harmonic currents, i_{hS}, will show up on the source side. It can be shown that the current drawn by the six-pulse bridge contains harmonics of the order

$$n = (P * i \pm 1) \tag{2.1}$$

where i is an integer greater than or equal to 1.

For a six-pulse converter, $P = 6$ and the line current contains harmonics of the order 5, 7, 11, 13, …. These are referred to as the characteristic harmonics of the six-pulse converter. For a 12-pulse converter (two 6-pulse units in parallel), its characteristic harmonics will be 11, 13, 23, 25, 35, 37, …. This is the reason behind the common practice to control harmonics by employing converters with higher numbers of pulses.

For a six-pulse converter, the following observations apply:

No triplen harmonics are present.

There are harmonics of order $6k \pm 1$ for integer values of k.

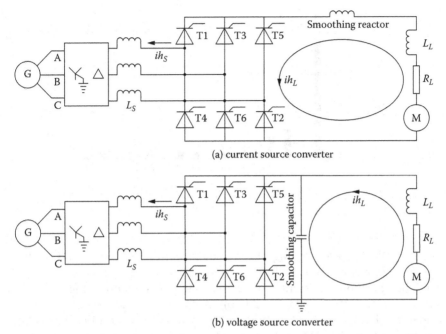

(a) current source converter

(b) voltage source converter

ih_S = Harmonic current due to commutation of thyristors. It may reach the AC source.

ih_L = Harmonic current produced in the rotating machine itself. It is confined to the load side.

FIGURE 2.11
Six-pulse converter used as a DC motor speed controller.

Harmonics of order $6k + 1$ are of positive sequence.

Harmonics of order $6k - 1$ are of negative sequence.

Twelve-pulse converters are powered from a three-winding (or phase shift) transformer, with a phase difference of 30° between the secondary and the tertiary; each connects to a converter's bridge. These converters create harmonics of order $12k$ (±) 1 at the source side. The harmonic currents of order $6k$ (±) 1 (with k odd), i.e., $k = 5, 7, 17, 19$, etc., flow between the secondary and tertiary of the phase shift transformer but do not make their way into the AC network.

The amplitude of the harmonic current on the converter front end will be influenced by the presence of a smoothing reactor such as that shown in Figure 2.11(a). For a six-pulse diode bridge having a large smoothing reactor, the magnitude of the harmonics can be approximated by the expression

$$|I_h| \approx \frac{|I_{\text{fund}}|}{h} \tag{2.2}$$

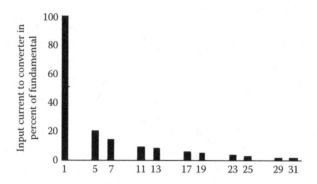

FIGURE 2.12
Harmonic spectrum of a six-pulse converter.

where I_h and I_{fund} are the magnitude of the nth-order harmonic and the fundamental current, respectively.

Higher harmonic currents can be expected if the smoothing reactor is small or nonexistent (see Figure 2.11(b)). Triplen harmonics can appear under unbalanced voltages. For this bridge, the harmonic spectrum will thus look like that in Figure 2.12. Note that the fifth harmonic shows a value of 20%, as obtained from Equation (2.2).

Fundamental mmf rotates in the positive direction, mmf from triplen harmonics is absent, and any fifth and seventh harmonic mmfs rotate in the negative and positive directions, respectively. Thus, from looking at the spectrum of Figure 2.12, it is possible to understand that negative sequence torques (from 5th, 11th, 17th, etc., harmonics) will be strongly interacting with positive sequence torques (from 7th, 13th, 19th, etc., harmonics) to produce torsional pulsating torques. This may explain the increased vibration levels sometimes experienced in applications involving synchronous generators feeding large VFD industrial applications, as further discussed in Chapter 4.

From Alex McEachern's *Teaching Toy*, Edition 2.0 (a useful and educational free harmonics tool particularly suitable for students),[8] a three-phase bridge rectifier would reveal the current waveform depicted in Figure 2.13.

2.3.3.1 Large Power Converters

These are used in electric utility applications in which large blocks of energy are transformed from AC to DC. Their nominal power is in the megavoltampere range, and generally, they present a much higher inductance on the DC than on the AC side. Therefore, the DC is practically constant and the converter acts as a harmonic voltage source on the DC side and as a harmonic current source on the AC side. Furthermore, in a perfectly balanced system, all resultant currents are the same in all phases.

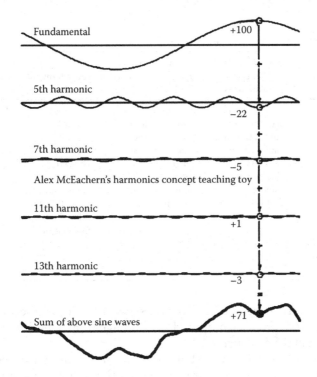

Fundamental +100

5th harmonic −22

7th harmonic −5

Alex McEachern's harmonics concept teaching toy

11th harmonic +1

13th harmonic −3

Sum of above sine waves +71

FIGURE 2.13
Three-phase rectifier. (Adapted from McEachern, A., *Power Quality Teaching Toy*, Edition 2.0, 2005.)

As described in Chapters 10 and 11, the worldwide massive integration of renewable solar and wind power taking place in industry is likely to represent a sizable amount of the energy consumed in the world. Wind farms sized 100 to 200 MW will be widespread, and their integration to the power utility grid will represent one of the most common large power converters.

2.3.3.2 Medium-Size Power Converters

Kilovoltampere-size converters are in this range and are found in increasing numbers in industry. The first applications in the industry were for DC motor speed control that still represents the major market for these types of converters. However, they are increasingly used in AC induction motor speed control. Many applications are now seen at land and offshore applications in the oil industry, where submersible pumping systems are used with variable frequency drives as artificial methods for oil production. Furthermore, the advent of power transistors and GTO thyristors is progressively stimulating the use of power converters for AC motor speed control.

TABLE 2.1

Some Power Converter Applications in Industry

Manufacturing Industry	Paper and Steel Industries	Transportation, Textile, and Food Industries	Petrochemical Industry	Residential Installations
Agitators, pumps, fans, and compressors in the process industry Planers, winches, drill presses, sanders, extruders, grinders, mills, and presses in machining	Blowers and compressors in heating and air conditioning Hoists and steel mill rollers	Elevators, trains, automobiles in transportation Looms in textiles Sheers in packaging Conveyors and fans in food industries	Compressors, variable frequency drives to power oil pumps, fans, cranes, and shovels in the oil and gas installations	Heat pumps, freezers, and washing machines

Similar to large-size power converters, the fifth harmonic in medium-sized converters can reach amplitudes that range from one-fifth to one-third the fundamental rated current.

In the case of electric railroad applications, it is common to see individual controls in every rectifier bridge. During the initial accelerating period with maximum current in the DC motor, the rectifier bridge produces the worst harmonic currents and operates at a low power factor. To alleviate this condition at low speeds, one of the bridges is bypassed while phase control is applied to the other bridge. Table 2.1 recaps the different applications of medium-size power converters.

2.3.3.3 Low-Power Converters

Uninterruptible power supplies (UPSs), welders, and printers are among these low-kilovoltampere-size power converter applications. It is common to see large commercial and public office buildings stuffed with computers and other peripheral devices. If they are additionally provided with UPSs to handle voltage sags and power supply interruptions, the amounts of harmonic currents can substantially increase. Residential areas at specific times of the day act as fabulous harmonic sources produced by all kinds of entertaining devices, as described previously.

The individual harmonics generated by battery charger circuits depend on the initial battery voltage. The overall harmonic content varies as a function of time and involves a random probability.

As in other appliances that use DC (TV sets, radio and stereo amplifiers, etc.), battery chargers produce zero-sequence harmonics, which overload the neutral conductor of the three-phase distribution transformer that supplies the single-phase, low-voltage loads. This is because the phase angle of the third harmonic does not vary enough to produce harmonic cancellation, so they are added up algebraically. As later discussed, fluorescent lighting also produces triplen harmonics, for which a concurrent use of battery chargers and fluorescent lamps from the same circuit can make things even worse.

Unlike the types of loads described earlier, whose nominal power is large enough to deserve an individual treatment, the loads we refer to in this section are important only when they represent a significant portion of the total load under concurrent operation. The Monte Carlo method can be used in some applications to investigate the probability of exceeding preset levels of harmonics from TV sets as well as from electric vehicle battery chargers serving multiple locations within the network.

2.3.3.4 Variable Frequency Drives

VFDs are, in reality, power converters. The reason to further address them under a separate section is because, by themselves, VFDs constitute a broad area of application used in diverse and multiple industrial processes. In a very general context, two types of VFDs can be distinguished: those that rectify AC power and convert it back into AC power at variable frequency and those that rectify AC power and directly feed it to DC motors in a number of industrial applications.

In both cases, the front-end rectifier, which can make use of diodes, thyristors, IGBTs, or any other semiconductor switch, carries out the commutation process in which current is transferred from one phase to the other. This demand of current "in slices" produces significant current distortion and voltage notching right on the source side, i.e., at the point of common coupling. Motor speed variations, which are achieved through firing angle control, will provide different levels of harmonic content on the current and voltage waveforms.

Variable frequency drive designs also determine where harmonic currents will predominantly have an impact. For example, voltage source inverters produce complex waveforms showing significant harmonic distortion on the voltage and less on the current waveforms. On the other hand, current source inverters produce current waveforms with considerable harmonic contents with voltage waveforms closer to sinusoidal. None of the drive systems is expected to show large distortion on both voltage and current waveforms, in line with Finney's observations.[7]

2.3.3.4.1 Distribution Static Compensator (DSTATCOM)

The DSTATCOM is a good example of a voltage source inverter (VSI) power electronics device connected in shunt to the distribution network. This is a concept imported from the application of FACTS (flexible AC transmission systems) comprehensively described by Hingorani and Gyugyi.[9] Among the objectives of the DSTATCOM are to eliminate harmonics from the power supply and to provide voltage and reactive power support during faults in the system. However, because the DSTATCOM uses a rectification bridge, a continuous harmonic production is created on the source side. Thus, the example presented here illustrates the harmonic voltage distortion in a circuit involving a distribution static compensator during and after a three-phase to ground fault.

The fault is simulated to occur at $t = 1.5$ s and last 0.75 s. The example, which is modeled using the PSCAD student edition software, involves a voltage control with proportional-integral (PI) controller and a pulse width modulation (PWM) controller, with a carrier frequency of nine times the fundamental and varying DC voltage. Figure 2.14 depicts the diagram of the six-pulse STATCOM setup, and Figure 2.15 shows the calculated voltage waveforms at both ends of the rectifying bridge in a time window that encompasses the start and end of the fault.

Figure 2.16 presents results of the simulation, which shows harmonic distortion and spectral content at the onset and extinction of the staged fault.

The THD_V levels found for voltage, Vna, during the transition times are as follows:

1.5 to 1.6 s: $THD_V = 12.77\%$ (fault starting period).

2.25 to 2.3 s: $THD_V = 13.95\%$ (fault clearing period)

2.5 s onward (until 2.5 s): $THD_V = 5\%$ (post-fault or steady state).

Notice that the total harmonic distortion levels during and at the clearing periods of the fault are more than twice the steady-state levels. At first glance, these levels are above the recommended standard values described in Chapter 3. However, standards do not cover harmonic distortion during transient conditions or during short-circuit faults. The fault example is chosen here to illustrate the ability of the software to calculate harmonic distribution change in rapid succession.

2.3.4 Fluorescent Lamps

Fluorescent tubes are highly nonlinear in their operation and give rise to odd harmonic currents of important magnitude. As a brief portrayal of the fluorescent lamp operation, we can state that magnetic core inductors or chokes contained inside the start ballasts function to limit the current to the tube. Likewise, they use a capacitor to increase the efficiency of the ballast by increasing its power factor. Electronic ballasts operate at higher frequency, which permits the use of smaller reactors and capacitors. The use of higher

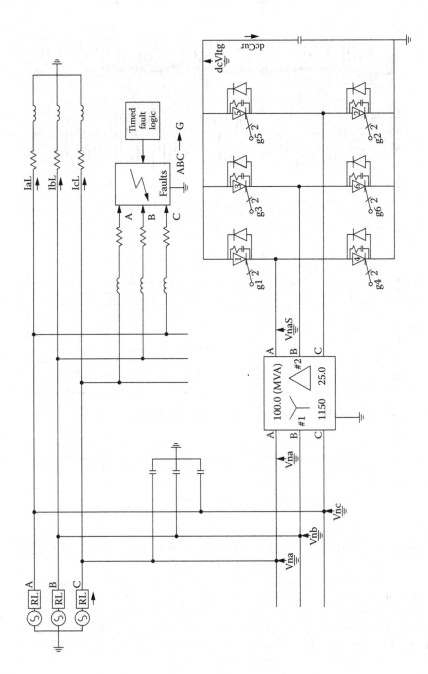

FIGURE 2.14
Six-pulse STATCOM.

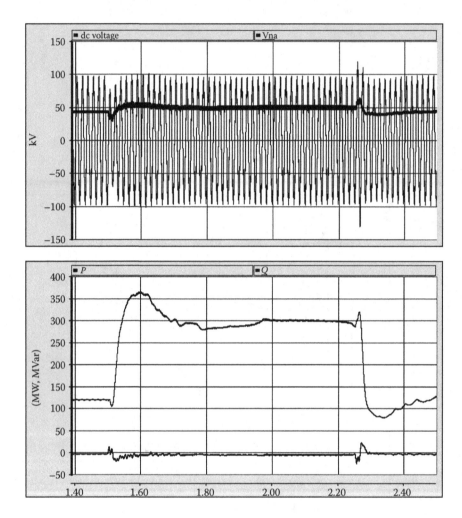

FIGURE 2.15
Voltage waveforms at both sides of the DSTATCOM converter.

frequencies allows them to create more light for the same power input. This is advantageously used to reduce the input power.

In a four-wire, three-phase load, the dominant phase current harmonics of fluorescent lighting are the third, fifth, and seventh if they use a magnetic ballast and the fifth with an electronic ballast (as adapted from Tolbert et al.[10] and presented in Figure 2.17). Triplen harmonics are added in the neutral, being the third dominant for a magnetic ballast, but multiple harmonics if an electronic ballast is used. See Figure 2.18.

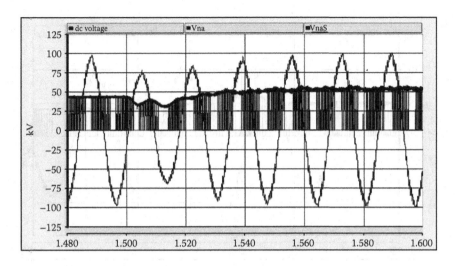

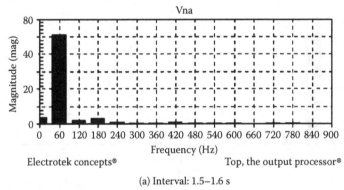

(a) Interval: 1.5–1.6 s

FIGURE 2.16
Harmonic spectra of the PWM voltage control system of Figure 2.15.

In Figure 2.18, notice the impressive amount of third-order harmonics in the neutral, particularly because they are all added in phase. It will be nonetheless important to remember that the current in the neutral must be determined from

$$\text{Ineutral} = \sqrt{I_1^2 + I_3^2 + I_5^2 + I_7^2 + I_9^2 + I_{11}^2 + \dots + I_n^2} \tag{2.3}$$

Furthermore, lighting circuits frequently involve long distances and combine with a poorly diversified load. With individual power factor correction capacitors, the complex LC circuit can approach a resonant condition around

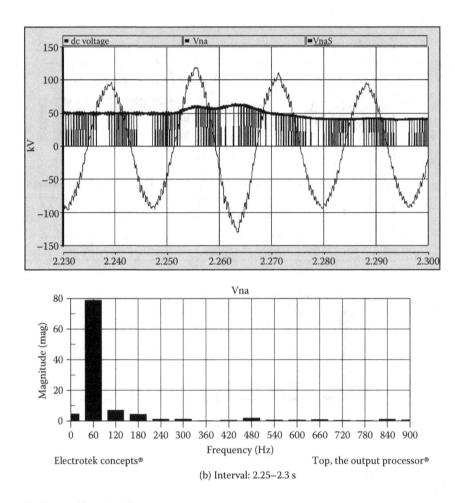

FIGURE 2.16 (Continued)
Harmonic spectra of the PWM voltage control system of Figure 2.15. *(Continued)*

the third harmonic. Therefore, these are significant enough reasons to over-size neutral wire lead connections in transformers that feed installations with substantial amounts of fluorescent lighting. Capacitor banks may be located adjacent to other loads and not necessarily as individual power factor compensators at every lamp.

2.3.5 Electric Furnaces

The melting process in industrial electric furnaces is known to produce substantial amounts of harmonic distortion. The introduction of fundamental frequency harmonics develops from a combination of the delay in the

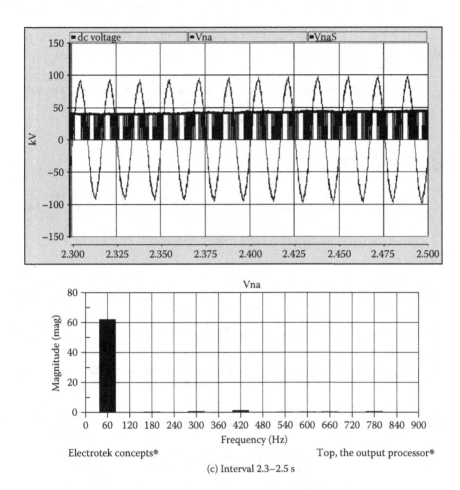

FIGURE 2.16 (Continued)
Harmonic spectra of the PWM voltage control system of Figure 2.15. *(Continued)*

ignition of the electric arc along with its highly nonlinear voltage-current character. Additionally, voltage changes caused by the random variations of the arc give rise to a series of frequency variations in the range of 0.1 to 30 kHz; each has its associated harmonics. This effect is more evident in the melting phase during the interaction of the electromagnetic forces among the arcs. Figure 2.19 shows plots of (a) the electric furnace power and (b) the current harmonics without attenuation of harmonic filters in a typical electric furnace application. The example includes the application of the Smart Predictive Line Controller, a patented Hatch technology[11] for arc stabilization and flicker reduction on AC electric arc furnaces.

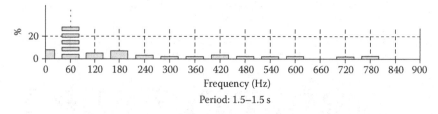

Frequency (Hz)

Period: 1.5–1.5 s

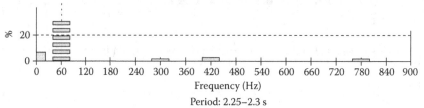

Frequency (Hz)

Period: 2.25–2.3 s

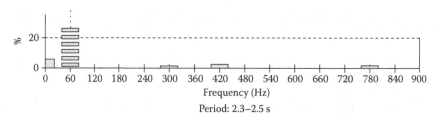

Frequency (Hz)

Period: 2.3–2.5 s

(d) Summary of harmonic spectra including commencement,
end- and postfault periods, respectively

FIGURE 2.16 (Continued)
Harmonic spectra of the PWM voltage control system of Figure 2.15.

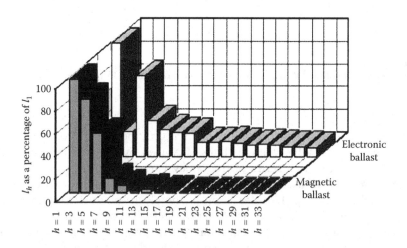

FIGURE 2.17
Harmonic spectra of fluorescent lamps for phase currents. (Data from Tolbert, L.M., Survey of Harmonics Measurements in Electrical Distribution Systems, in *IEEE IAS Annual Meeting*, San Diego, California, October 6–10, 1996, 2333–2339.)

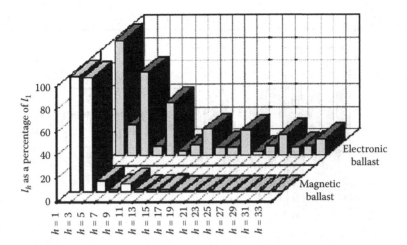

FIGURE 2.18
Harmonic spectra of fluorescent lamps for neutral currents. (Data from Tolbert, L.M., Survey of Harmonics Measurements in Electrical Distribution Systems, in *IEEE IAS Annual Meeting*, San Diego, California, October 6–10, 1996, 2333–2339.)

2.4 Future Sources of Harmonics

The challenge for electrical system designers in utilities and industry is to design the new systems or adapt the present systems to operate in environments with escalating harmonic levels. The sources of harmonics in the electrical system of the future will be diverse and more numerous. The problem becomes complicated with the increased use of sensitive electronics in industrial automated processes, personal computers, digital communications, battery charges, and multimedia.

Utilities, which generally are not regarded as large generators of harmonics, may be lining up to join current harmonic producers with the integration of distributed resources on the rise. Photovoltaic, wind, natural gas, carbonate full cells, and even hydrogen are expected to play increasingly important roles in managing the electricity needs of the future. Distributed generators that presently provide support to utilities, especially during peak demand hours, will be joined by numerous harmonic-producing units, fueled by natural gas or even wind, in the so-called microturbine concept.

Chapters 10 through 12 are intended to provide a more comprehensive review of harmonics to be expected from integration of wind and power renewable resources, smart grid devices, HVDC, and battery chargers in the vehicle to grid concept.

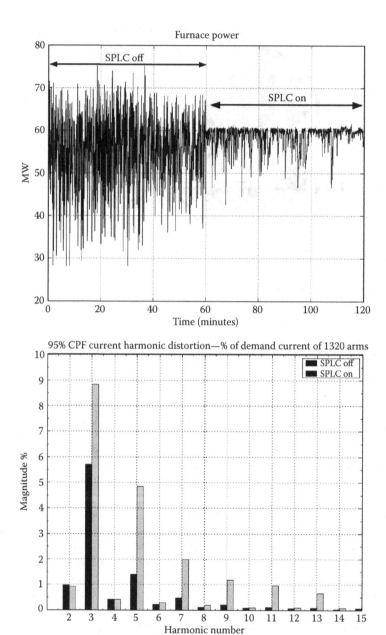

FIGURE 2.19
Furnace power and harmonic spectrum drawn at a 60 MW electric furnace by a typical arc furnace. (Adapted from private communication with Hatch, Smart Predictive Line Controller, 2005.)

References

1. Power System Computer Aided Design (PSCAD), http://pscad.com.
2. IEEE 519-1992, *Recommended Practices and Requirements for Harmonic Control in Electric Power Systems*.
3. IEEE WG on Modeling and Analysis of System Transients Using Digital Programs, *IEEE Trans. Power Delivery*, 15(1), 2000.
4. Ferrucci, P., Ferroresonance, *Cashier Tech.*, Schneider No. 190, ETC 190, March 1998.
5. Enrigth, W., Watson, N., and Nayak, O.B., Three Phase Five-Limb Unified Magnetic Equivalent Circuit Transformer Models for PSCAD V3, in *IPST '99 Proceedings*, Budapest, 1999, 462–467.
6. Enrigth, W., Nayak, O.B., Irwin, G.D., and Arrillaga, J., An Electromagnetic Transient Model of the Multi-Limb Transformers Using Normalized Core Concept, in *IPST '97 Proceedings*, 1997, 93–108.
7. Finney, D., *Variable Frequency AC Motor Drive Systems*, Peter Peregrin Ltd., on behalf of the Institution of Electrical Engineers, IEE Power Engineering Series 8, London, 1988.
8. McEachern, A., *Power Quality Teaching Toy*, Edition 2.0, 2005.
9. Hingorani, N.G., and Gyugyi, L., *Understanding FACTS: Concepts and Technology of Flexible AC Transmission Systems*, IEEE Press, New York, 1999.
10. Tolbert, L.M., Survey of Harmonics Measurements in Electrical Distribution Systems, in *IEEE IAS Annual Meeting*, San Diego, October 6–10, 1996, 2333–2339.
11. Private communication from Hatch, Smart Predictive Line Controller, 2005.

3

Standardization of Harmonic Levels

3.1 Introduction

The most widespread standards for harmonic control worldwide are due to the Institute of Electrical and Electronics Engineers (IEEE) in the United States and the International Electrotechnical Commission (IEC) in the European Union. In 1981, the IEEE issued standard 519-1981,[1] which aimed to provide guidelines and recommended practices for commutation notching, voltage distortion, telephone influence, and flicker limits produced by power converters. The standard contended with cumulative effects but did little to consider the strong interaction between harmonic producers and power system operation.

The main focus of the revised IEEE 519 standard in 1992[2] was a more suitable stance in which limitations on customers regarding maximum amount of harmonic currents at the connection point with the power utility did not pose a threat for excessive waveform distortion. This revision also implied a commitment by power utilities to verify that any remedial measures taken by customers to reduce harmonic injection into the distribution system would reduce the voltage distortion to tolerable limits. The interrelation of these criteria shows that the harmonic problem is a system, and not a site, problem. Compliance with this standard requires verification of harmonic limits at the interface between utilities and customers, more commonly known as point of common coupling (PCC).

Recommended total harmonic distortion (THD) levels for current and voltage signals were established in the 1992 revision of IEEE 519. Total and individual harmonic distortion levels were issued for customers on current and for utilities on voltage signals at the PCC. The total demand distortion (TDD) concept was created to better relate the THD to the demand current. The TDD is the total root sum square (RSS) of the harmonic current expressed in percentage of the nominal maximum demand load current. The standard also called for limiting commutation notching levels at individual low-voltage customer locations. Communication interference with systems produced by low-voltage DC converters was addressed in the revised standard, and IT limits for utilities were also established. All recommended IEEE

519-1992 limits were presented for different voltage levels encompassing 69 kV and below, 69.001 through 161 kV, and higher than 161 kV.

The 1992 edition of the IEEE harmonic standard thus advocates the joint contribution of utilities and customers to contend with harmonic emission and control matters. The compliance with recommended levels was deemed a convenient way to keep harmonic current penetration into the distribution system under control and permit the operation of equipment and devices that draw current in a nonlinear fashion. For over a decade, this standard has been the main reference for American utilities, customers, and manufacturers alike in trying to sustain the operation of the electrical systems within tolerable waveform distortion margins. Harmonic distortion limits were not specified for particular types of customers (industrial, commercial, or residential). The observance of these limits also brought the need for properly modeling utility systems and harmonic sources, which led to the development of expert software whose main characteristics are described in Chapter 8.

Similar to IEEE 519, IEC harmonic standards set limits at the utility-customer interface; they also set limits for customer equipment, in a clear reference to residential installations. After multiple revisions, the last IEC harmonic standard 61000-3-2[3] focused on limiting equipment consumption of harmonics. The equipment refers to single- and three-phase units with per-phase currents up to 16 A. Individual harmonic limits are required for every one of the four different classes of equipment considered—namely, A through D. Class D is regarded as a highly harmonic producer, and its harmonic content is subject to a strict maximum harmonic current per unit of the current drawn at the main frequency. These are more stringent limits than for the other classes of equipment.

Regarding voltage distortion, compatibility[4] and planning[5] levels are specified for electrical networks to tie in with emission and immunity levels, respectively, in low voltage (LV) and medium voltage (MV) installations. Compatibility levels are used as a reference for coordinating the emission and immunity of the equipment in LV and MV installations. Planning levels are used by system operators in evaluating the impact of all disturbing loads on the utility supply. For MV, compatibility levels are described in IEC 61000-2-12.[4] Indicative values for planning levels along with definitions of LV, MV, HV, and extra high voltage (EHV) are given in IEC 61000-3-6[5]: LV is 1 kV or less, MV is above 1 kV and below 35 kV, HV is from 35 to 230 kV, and EHV is above 230 kV.

Unlike IEEE 519, IEC considers the harmonic distortion assessment to cover short- and long-term effects.[5] The former are referred to as very short (3 s) events, and the latter as short-period (10 min) events. Very short-time events are meant to account for disturbing effects on electronic devices that may be susceptible to harmonic levels lasting up to 3 s, excluding transients. Long-term effects account for thermal effects on equipment such as transformers, motors, cables, capacitor banks, etc. However, for statistical assessment, periods of 1 week or longer are recommended.[5-8]

Interharmonic (harmonic components not an integer of the fundamental frequency) voltage compatibility limits related to flicker in lighting devices are addressed in IEC 61000-2-2.[7] IEEE 519-1992 does not specifically set limits for interharmonics. However, it is expected that this issue will be included in further revisions of the standards.[9,10]

3.2 Harmonic Distortion Limits

The rms value of a voltage waveform, considering the distortion produced by harmonic currents, is expressed by

$$Vrms = \sqrt{\sum_{h=1}^{\infty} Vh^2} \qquad (3.1)$$

Likewise, the rms value of a sinusoidal current, taking into account the distortion created by the harmonic source currents, is given by

$$Irms = \sqrt{\sum_{h=1}^{\infty} Ih^2} \qquad (3.2)$$

As defined in Chapter 1, total harmonic distortion is a parameter used in IEEE and IEC standards. For the sake of convenience, the definition of THD discussed in Chapter 1 is repeated here for voltage and current signals, respectively:

$$THD_V = \frac{\sqrt{\sum_{h=2}^{\infty} V_h^2}}{V_1} \qquad (3.3)$$

$$THD_I = \frac{\sqrt{\sum_{h=2}^{\infty} I_h^2}}{I_1} \qquad (3.4)$$

3.2.1 In Agreement with IEEE 519-1992

Per IEEE 519,[2] recommended harmonic distortion limits are to be verified through comparison with measurements at the PCC, i.e., the interface between the electric utility and the customer. Chapter 5 describes the

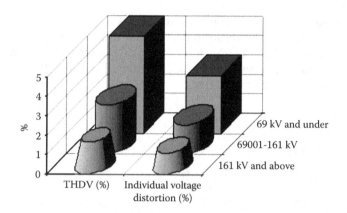

FIGURE 3.1
IEEE 519 voltage harmonic distortion limits.

relevant aspects involved in the measurements of harmonics. A significant issue is that levels can be exceeded by 50% under start-ups or unusual conditions with durations less than an hour. It additionally recommends the use of probability distribution functions from field measurements, stating that if limits are exceeded for only a short period, such a condition can be considered acceptable.

As portrayed in Figure 3.1, the recommended limits are a function of the system voltage level. For electric networks of 69 kV and below, for example, the total voltage distortion is limited to 5%; no individual voltage harmonic should exceed 3%, as depicted in the illustration.

Concerning current harmonic distortion, IEEE 519 defines the limits as a function of the ratio between the short-circuit current at the PCC (I_{sc}) and the average current corresponding to the maximum demand during a period of 12 months (I_L). The recommended limits are summarized in Figure 3.2. Notice that the suggested limits become more stringent for decreasing I_{sc}/I_L ratios and increasing harmonic order.

The following aspects are to be noticed:

Regardless of the I_{sc}/I_L ratio at the PCC, all power generation equipment must meet the values given for an I_{sc}/I_L ratio < 20.

Even harmonics are limited to 25% of the odd harmonic limits.

Current distortions that result in a DC offset (e.g., half-wave converters) are not allowed.

Note that total harmonic distortion limits are expressed in terms of the total demand distortion (TDD), which refers to the electric demand during a period of 15 to 30 min.

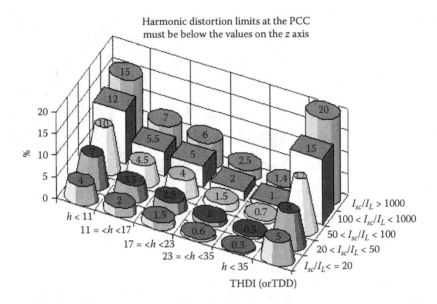

FIGURE 3.2
IEEE 519 current distortion limits.

The shown values are the maximum permissible limits under continuous operation. During start-up or unusual operating conditions lasting less than 1 h, these limits can be exceeded by 50%.

In systems that use converters of more than six pulses, these limits can be increased by $\sqrt{\frac{q}{6}}$, where q is the number of pulses.

IEEE 519 also sets limits for commutation notching in LV power converters. Figure 3.3 depicts the allowed notch depth and notch area along with the corresponding THD.

IEEE 519-1992 practically establishes a joint involvement of the customer and the electric company to maintain harmonic distortion levels within limits that will enable the power distribution systems to operate within safe voltage distortion limits. Making customers play a part in this process ensures the direct benefit of minimizing the negative effects that harmonic currents may have at their premises. Figure 3.4 outlines this relationship.

3.2.2 In Conformance with IEC Harmonic Distortion Limits

With regard to IEC, as described earlier, compatibility levels are used as a reference for coordinating the emission and immunity of the equipment in LV and MV installations. Planning levels are used by system operators in

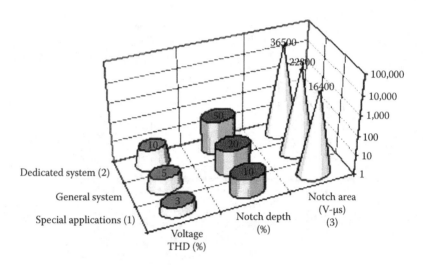

(1) Special applications include hospitals and airports
(2) A dedicated system is exclusively dedicated to converter loads
(3) In V-μs for rated voltage and current. If voltage is different than 480 V, multiply by V/480

FIGURE 3.3
Notching levels from IEEE 519. (Data from IEEE 519-1981, *IEEE Guide for Harmonic Control and Reactive Compensation of Static Power Converters.*)

evaluating the impact of all disturbing loads on the utility supply. For the interface utility-customer, short-time (10 min) measurement results of voltage distortion, usually taken as the value related to the 95% probability weekly value,[5] must conform to planning levels. Planning levels are defined in IEC 61000-2-12[4]; higher-level emissions reaching up to 11% for very short periods

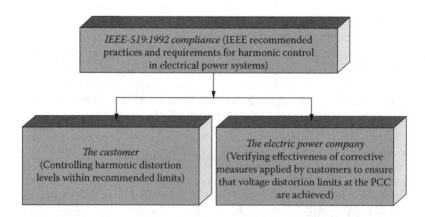

FIGURE 3.4
Relationship between customers and power utilities to achieve compliance with IEEE 519.

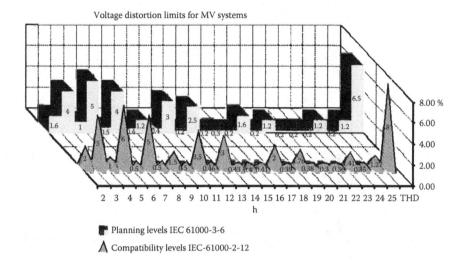

FIGURE 3.5
IEC voltage distortion limits.

(3 s) are also considered. These are important in assessing very short-time effects of harmonics.

Figure 3.5 shows the IEC compatibility[4] and planning[5] individual voltage distortion levels for MV systems. As noted, THD levels for MV systems are somewhat higher than those from IEEE in Figure 3.1.

Regarding customer equipment, IEC standards specifically set harmonic levels aimed at protecting low-voltage systems at customer and utility installations. IEC 61000-3-2[3] considers four different classes of equipment in establishing harmonic current limits:

Class A: Balanced three-phase equipment; household appliances (excluding equipment) identified as class D; tools (except portable), dimmers for incandescent lamp (but not other lighting equipment), audio equipment; anything not otherwise classified

Class B: Portable power tools

Class C: All lighting equipment except incandescent lamp dimmers

Class D: Single phase; under 600 W; personal computer, PC monitor, TV receiver

Table 3.1 presents the limits for individual harmonic current for every one of the classified equipment classes. Limits are given in amperes for equipment classes A and B and in percentage of fundamental for class C. For class D, levels are specified in milliamperes per watt for equipment with a rated power exceeding 75 W but inferior to 600 W, or in amperes for equipment larger than 600 W. Notice that total harmonic distortion levels are not

TABLE 3.1

IEC 61000-3-2 Harmonic Current Limits for Different Equipment Classes

Odd Harmonic n	Even Harmonic n	Max. Permissible Harmonic Current (A)		Max. Permissible Harmonic Current (% of fundamental)	Max. Permissible Harmonic Current (mA/W) 75 W < P < 600 W	Max. Permissible Harmonic Current (A) P > 600 W
		Class A	Class B	Class C	Class D	
	2	1.08	1.62	2		
3		2.3	3.45	(30) × circuit power factor	3.4	2.3
	4	0.43	0.645			
5		1.14	1.71	10	1.9	1.14
	6	0.3	0.45			
7		0.77	1.155	7	1	0.77
	$8 \leq n \leq 40$	$1.84/n$	$2.76/n$			
9		0.4	0.6	5	0.5	0.4
11		0.33	0.495	3 (for all $11 \leq n \leq 39$)	0.35	0.33
13		0.21	0.315		0.296	0.21
$15 \leq n \leq 39$		$2.25/n$	$3.375/n$		$3.85/n$	$2.25/n$

specified. For an application involving 230 V class D equipment subject to the maximum level of individual harmonic current, the total harmonic distortion would yield around 95%.[11]

Finally, IEC 61000-2-2[7] defines compatibility levels for situations of interharmonic voltages occurring near the fundamental frequency. Specific types of loads are sensitive to the square of the voltage and exhibit a beat effect resulting in flicker. Figure 2 in reference 7 describes maximum interharmonic amplitudes (as a percentage of fundamental voltage) as a function of the difference between the interharmonic and the fundamental frequency (beat frequency). In essence, voltage oscillations between 1 and 4% are established for beat frequencies below 1 Hz or between 20 and 40 Hz (too small or too large differences), and voltage variations between about 0.2 and 1% are set for beat frequencies between 1 and 25 Hz for 120 and 230 V lamps.

Other effects of interharmonics and subharmonics, including their detrimental effects on underfrequency relays and harmonic torques in rotating machines that are not addressed in the IEC standards, are described in reference 8.

References

1. IEEE 519-1981, *IEEE Guide for Harmonic Control and Reactive Compensation of Static Power Converters*.
2. IEEE 519-1992, *Recommended Practices and Requirements for Harmonic Control in Electric Power Systems*.
3. IEC 61000-3-2, *Electromagnetic Compatibility (EMC)—Part 3-2: Limits—Limits for Harmonic Current Emissions (Equipment Input Current ≤ 16 A per Phase)*, 2001–10.
4. IEC 61000-2-12, *Electromagnetic Compatibility (EMC)—Part 2-12: Compatibility Levels for Low-Frequency Conducted Disturbances and Signaling in Public Medium-Voltage Power Supply Systems*, 2003–04.
5. IEC 61000-3-6, *Assessment of Emission Limits for Distorting Loads in MV and HV Power Systems*, technical report type 3, 1996.
6. IEC 61000-4-30, *Power Quality Measurement Methods*, 2003.
7. IEC 61000-2-2, *Electromagnetic Compatibility—Part 2-2, Environment Compatibility Levels for Low-Frequency Conducted Disturbances and Signaling in Public and Low-Voltage Power Supply Systems*, 2002.
8. Joint WG CIGRE C4.07/CIRED, *Power Quality Indices and Objectives*, final WG report, January 2004, rev. March 2004.
9. Fuchs, E.F., Roesler, D.J., and Masoum, M.A.S., Are Harmonic Recommendations according to IEEE and to IEC Too Restrictive? *IEEE Trans. Power Delivery*, 19(4), 2004.
10. Halpin, M., Harmonic Modeling and Simulation Requirements for the Revised IEEE Standard 519-1992, in *2003 IEEE Power Engineering Society General Meeting Conference Proceedings*, Toronto, Ontario, Canada, July 13–17, 2003.
11. Ward, J., and Ward, D., Single Phase Harmonics, in *PSER EMI, Power Quality, and Safety Workshop*, April 18–19, 2002.

4

Effects of Harmonics on Distribution Systems

4.1 Introduction

Assuming that nonlinear loads keep a steady expansion as load increases, harmonics on distribution systems are expected to grow as electricity demand increases. Therefore, the expected effects of harmonics described in this chapter will show a corresponding increase with load.

Predictions in energy growth are comprehensively described in the 2014 Energy Outlook.[1] According to this forecast, in the period 2012–2014 electricity demand will grow by 20 to 41%, depending on whether a low or high economic growth rate, respectively, is assumed. In the same period, the increase in total delivered energy consumption in the industrial sector will be around 28%.

In the commercial sector energy will grow by 0.6% per year. In this sector, miscellaneous electric loads, including medical equipment, video displays, and many other devices, will experience a growth of around 21% in the period 2012–2040. The shift from desktop to portable computing devices, as well as reduction in processor power, will cause a decrease of personal computer equipment by 5.6% annually. On the other hand, an increased electricity use in data centers and servers by non-PC office equipment will see a rise of 2% per year.

The residential sector is projected to experience a modest increase of 0.2% per year from 2010 to 2040.[1] A significant potential harmonic source that will have to be looked at is renewable energy, mainly solar and wind, constituting around 62.3% of commercial capacity by 2040. Because of falling prices for photovoltaic inverters and panels due to federal investment tax credits, solar power will see an increase of 5.7%.

Similarly, wind power capacity will see a growth of around 11.1% by 2040 largely due to federal and local incentives.

Overall, energy consumption is anticipated to see a decline in the nonmanufacturing subsector by 10% from 2012 to 2040, with both construction and agriculture showing a decline in energy of 17%, while the mining industry will undergo an increase in energy of 26%.

Considering the trend in electricity consumption and the increasing sources of harmonics in industrial, commercial, and residential installations over the next 25 years, in that order, it shall be necessary to clearly assess the effects that unfiltered harmonics from scattered nodes in the power system will have in different equipment, on operation and on diverse processes.

4.2 Thermal Effects on Transformers

Modern industrial and commercial networks are increasingly influenced by significant amounts of harmonic currents produced by a variety of nonlinear loads like variable speed drives, electric and induction furnaces, and fluorescent lighting. Add to the list uninterruptible power supplies and massive numbers of home entertaining devices, including personal computers.

All of these currents are sourced through service transformers. A particular aspect of transformers is that, under saturation conditions, they become a source of harmonics. Delta-wye- or delta–delta-connected transformers trap zero-sequence currents that would otherwise overheat neutral conductors. The circulating currents in the delta increase the rms value of the current and produce additional heat. This is an important aspect to watch out for. Currents measured on the high-voltage side of a delta-connected transformer may not reflect the zero-sequence currents, but their effect in producing heat losses is there.

In general, harmonics losses occur from increased heat dissipation in the windings and skin effect; both are a function of the square of the rms current, as well as of eddy currents and core losses. This extra heat can have a significant impact in reducing the operating life of the transformer insulation. Transformers are a particular case of power equipment that has experienced an evolution that allows them to operate in electrical environments with considerable harmonic distortion. This is the K-type transformer. Because losses and K-type transformers are further described in Chapter 9, here we only stress the importance of harmonic currents in preventing conventional transformer designs from operating at rated power under particular harmonic environments. In industry applications in which transformers are primarily loaded with nonlinear loads, continuous operation at or above rated power can impose a high operating temperature, which can have a significant impact on their lifetime.

4.2.1 Neutral Conductor Overloading

In single-phase circuits, return currents carrying significant amounts of harmonic components flow through transformer neutral connections increasing the rms current. Furthermore, zero-sequence currents (odd integer multiples of 3) add in phase in the neutral. Therefore, the operation of transformers in harmonic environments demands that neutral currents be evaluated in grounded-wye-connected transformers to avoid the possibility of missing the grounding connection as a consequence of overloading. In balanced three-phase, four-wire systems, there is no current on the neutral, for which the presence of neutral currents under these conditions should be attributed to the circulation of zero-sequence harmonics, which are mostly produced by single-phase power supplies.

In systems that are not entirely balanced, the unbalanced current circulates on the return (neutral) conductor. Because this conductor is usually sized the same as the phase conductors so that it can comfortably handle unbalanced currents, overheating may occur if those currents are subsequently amplified by zero-sequence currents. Large numbers of computers in office buildings make up a formidable source of harmonic currents produced by their electronic switched power supplies.

A common practice is to size neutral conductors to carry as much as two times the rated rms current of phase conductors. Monitoring temperature increase on the neutral conductor of transformers might be a good start to detect whether zero-sequence harmonic currents are not overstressing neutral connections. This is true as long as the system does not incur increased levels of current unbalance that would produce a rise in neutral conductor temperature.

4.3 Miscellaneous Effects on Capacitor Banks

4.3.1 Overstressing

Increased voltage can overstress and shorten the life of capacitor banks. Voltage, temperature, and current stresses are the drivers of capacitor bank conditions that lead to dielectric breakdown. The output reactive power from a capacitor bank varies with the square of the voltage, as described by

$$\text{VAR} = \frac{V^2}{X_c} \tag{4.1}$$

Operating voltage can increase in distribution systems under light load conditions or when fuse links operate to isolate a failed capacitor unit, leaving the remaining units exposed to an overvoltage condition. For example,

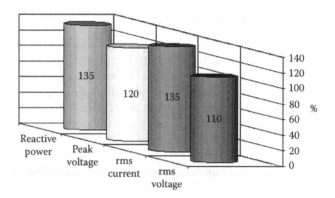

FIGURE 4.1
Allowed operation limits for shunt capacitor banks. (Data from IEEE 18-2002, *IEEE Standard for Shunt Power Capacitors.*)

a 5% increase in the nominal voltage of a capacitor unit would cause it to deliver $(1.05)^2 = 1.1$ or 110% of rated reactive power. Harmonic distortion is definitely another factor that contributes to impose voltage stresses on capacitor banks. This is a serious condition in industrial facilities with unfiltered large power converters.

IEEE 18[2] allows for the increase of fundamental operating parameters in capacitor banks as illustrated in Figure 4.1. The y-axis indicates values in percentage of rated values. These operating limits are meant for continuous operation. Thus, it will be important to take into account these limits also in the design of harmonic filters because capacitor banks in single-tuned filters are intended to act as a sink for the bulk of harmonic currents of the corresponding tuned frequency.

4.3.2 Resonant Conditions

As further addressed in Chapter 6, resonant conditions involve the reactance of a capacitor bank that at some point in frequency equals the inductive reactance of the distribution system, which has an opposite polarity. These two elements combine to produce series or parallel resonance. In the case of series resonance, the total impedance at the resonance frequency is reduced exclusively to the resistive circuit component. If this component is small, large values of current at such frequency will be developed. In the case of parallel resonance, the total impedance at the resonant frequency is very large (theoretically tending to infinite). This condition may produce a large overvoltage between the parallel-connected elements, even under small harmonic currents. Therefore, parallel resonant conditions may represent a hazard for solid insulation in cables and transformer windings and for the capacitor bank, and their protective devices as well.

Resonant frequencies can be anticipated if the short-circuit current level at the point where the capacitor bank is installed is known, following Equation (4.2):

$$h_r = \sqrt{\frac{kVA_{short\_circuit}}{kVAR_{cap\_bank}}} \tag{4.2}$$

where h_r is the resonant frequency as a multiple of the fundamental frequency, $kVA_{short\_circuit}$ is the short-circuit power available at the site, and $kVAR_{cap\_bank}$ is the reactive power rating of the capacitor bank.

Note how changing any of these parameters can shift the resonant frequency. This is a practice actually used sometimes in certain applications involving excessive heating in transformers connected to nonlinear loads.

If this resonant frequency coincides with a characteristic harmonic present at the site, that current will see a large upstream impedance and the existing voltage harmonic distortion will be amplified. Wagner et al.[3] suggested that capacitor banks can be applied without concern for resonance conditions as long as the nonlinear load and capacitor bank are less than 30 and 20%, respectively, the rated kilovoltamperes of the transformer, assuming a typical transformer impedance of around 5 to 6%. Otherwise, the capacitors should be used as a harmonic filter, with a series reactor that tunes them to one of the characteristic harmonics of the load. Generally, fifth and seventh harmonics are the most commonly found and account for the largest harmonic currents.

4.3.3 Unexpected Fuse Operation

As mentioned earlier, rms voltage and current values may increase under harmonic distortion. This can produce undesired operation of fuses in capacitor banks or in lateral feeding industrial facilities that operate large nonlinear loads. Capacitor banks can be further stressed under the operation of a fuse on one of the phases, which leaves the remaining units connected across the other phases. They are thus left subject to an unbalanced voltage condition that can produce overvoltages and detune passive harmonic filters if they are not provided with an unbalance detection feature.

An important aspect to look after with the advent of cogeneration and microturbine schemes using power inverters with electronic switching technology will be their harmonic current contribution and how this may affect the operation performance of islanding protective relays.

4.4 Abnormal Operation of Electronic Relays

Variable frequency drive (VFD) operation leading to shut-down conditions is often experienced in applications involving oil fields in which solid material (sand) abruptly demands higher thrust power, mining works in which sudden increases in lifting power occur, and high inertia loads, among others. In all these cases, the protective relays trip as a response to overcurrents exceeding the established settings. Similar effects can be experienced under the swift appearance of harmonic distortion on current or voltage waveforms exceeding peak or rms preset thresholds. Therefore, when protective relays trigger during the operation of a nonlinear load, harmonic distortion should be assessed. It might well be that an unpredicted overloading condition is the cause of the unexpected operation, but often increased harmonic levels following nonlinear load growth are the reason for similar behavior.

On the other hand, third harmonic currents produced by severe line current unbalance may cause nuisance relay tripping in VFD applications. Therefore, nuisance and missed relay tripping in installations with nonlinear loads should be assessed by checking the harmonic distortion levels and by inspecting the relays for possible threshold-setting fine-tuning. The onset of this type of occurrence in industrial installations may be used as a warning to start considering harmonic filtering actions.

4.5 Lighting Devices

Chapter 2 presented some examples of harmonic current generation in fluorescent lamps using magnetic and electronic ballasts. This phenomenon, though, does not produce a self-impact on lighting luminosity levels. It appears that frequency components that are a noninteger multiple of the fundamental frequency, also called interharmonics, are more prone to excite voltage oscillations that lead to light flickering. The main sources of interharmonics are the cycloconverters widely used in the steel, cement, and mining industries, as well as arc welders and furnaces. According to the joint IEEE Task Force and CIGRE/CIRED Working Group on Interharmonics,[4] and to reference 5, cycloconverters have characteristic frequencies of

$$f_i = (p_1 \cdot m \pm 1) \cdot f_1 \pm p_2 \cdot n \cdot f_0 \qquad (4.3)$$

where f_i is the interharmonic characteristic frequency, f_1 is the fundamental frequency, p_1 and p_2 are number of pulses on the rectifier and output sections, respectively, m and n are integers 0, 1, 2, 3, ..., but not 0, at the same time, and f_0 is output frequency of the cycloconverter.

Light flicker is one of the main impacts of interharmonics due to the modulated steady-state interharmonic voltage on the power frequency voltage.

According to reference 4, the rms voltage fluctuations that can be produced by interharmonic phenomena can be expressed by

$$U = \sqrt{\frac{1}{T}\int_0^T [\sin 2\pi f_1 t) + a\sin(2\pi f_i t)^2]dt} \qquad (4.4)$$

where a is the amplitude of the interharmonic voltage in per unit (p.u.) with $a = 1$ for the fundamental frequency.

It should be noted that incandescent lamps are more sensitive to flickering responding to rms voltage variations, and fluorescent lamps are more sensitive to peak voltage fluctuations.

4.6 Telephone Interference

The common construction of telephone lines built underneath power conductors on electric utility distribution poles makes them prone to a number of interference phenomena. Arrillaga et al.[6] describe the inductive, capacitive, and conductive interference that can take place between a power and a telephone line. In Chapter 1, the telephone influence factor (TIF) and the IT product were described as some of the power quality indices used by IEEE 519[7] to address and recommend limits on the telephone interference issue under harmonic distortion. According to reference 7, an IT product over 25,000 will probably cause interference problems.

4.7 Thermal Effects on Rotating Machines

Similar to transformers, rotating machines are exposed to thermal effects from harmonics. Because the effective resistance of a conductor goes up as frequency rises, a current wave rich in harmonics may cause greater heating on winding conductors than a sine wave of the same rms value. The overall effect can lead to a decreased transformer lifetime. The most significant aspects of rotating machine losses due to harmonics are described in Chapter 9.

4.8 Pulsating Torques in Rotating Machines

Additionally, magnetomotive forces (mmfs) induced by positive and negative-sequence harmonics interact with the nominal frequency mmf force, creating torque components of different frequencies (as described by Escobar and De La Rosa[8]). This may lead to problems on the shaft of rotating machines subject to the influence of harmonic torsional pairs, including

Equipment fatigue

Unexplained operation of mechanical fuses (bolts used to bond together turbine and generator shafts)

Increased vibration

Bearing wear-out

As an illustration, we analyze the case of a turbine-generator set whose mechanical shaft model is illustrated in Figure 4.2.[8]

The equation of motion of a mechanical system describing the developed torque as a function of angular displacement and inertia constant, J, can be described per

$$J\ddot{\theta} + \dot{D} + K\theta = T \tag{4.5}$$

By disregarding the damping matrix, through modal analysis we can describe the eigenvalues (natural frequencies of the system) as follows:

$$\Delta = |K - J\omega^2 I| = 0 \tag{4.6}$$

$$\lambda = |\omega_1^2 \omega_2^2 \omega_3^2 \cdots \omega_i^2| \tag{4.7}$$

Assuming that matrix D is composed of viscous damping and that it has the form of a linear combination of matrices J and K, we can write

$$\theta = Xq \tag{4.8}$$

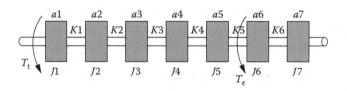

FIGURE 4.2
Model of a mechanical turbine-generator system.

where the eigenmatrix X includes eigenvectors of every mode and q is the vector of the new variables.

By using orthogonality properties of eigenvectors, we can uncouple the equation of motion:

$$J^{-1}(X^t J X \ddot{q} + X^t D X \dot{q} + X^t K X q = X^t T \tag{4.9}$$

which leads to n decoupled second-order differential equations of the form

$$\ddot{q} + 2\zeta_i \, w_i \dot{q} + \omega_i^2 q = Q_i \tag{4.10}$$

We integrate this equation to get the simulation in the time domain.

Let us now look at the steady-state amplitude of the electric torque polluted with one oscillating component. As input, we have one pulsating component of electrical torque as

$$T_e = T_b + T_h \sin(\omega_f t) \tag{4.11}$$

where ω_f is the frequency of the pulsating torque. Then Q_i has the form:

$$Q_i = C_{1i} + C_{2i} \sin(\omega_f t) \tag{4.12}$$

The solution for a steady-state sinusoidal excitation torque (Q_i) for the uncoupled differential equation is

$$q_i(t) = \frac{C_{1i}}{\omega_i^2} + |\vec{q}| \sin(\omega_f t - \varphi) \tag{4.13}$$

where phasor $^{TM}q^{TM}$ is

$$\vec{q} = \frac{c_{21}}{\omega_i^2 \sqrt{\left(1 - \frac{\omega_f^2}{\omega_i^2}\right)^2 + \left(2\zeta_i \frac{\omega_f}{\omega_i}\right)^2}} \tag{4.14}$$

As an example, let us assume a single turbine-generator set subject to a harmonic load around 16% the capacity of the generator. This load is connected to the generator bus through a 5 MVA transformer, as illustrated in Figure 4.3, where the assumed natural frequencies are also indicated.

Using the preceding analytical approach, we calculate the electric torque for the generator using the harmonic spectrum of the VFD converter and obtain the results presented in Table 4.1 and in Figure 4.4, in which the electric torque is plotted together with the harmonic spectrum of the VFD converter. Note the occurrence of electric torques at the intermediate frequencies

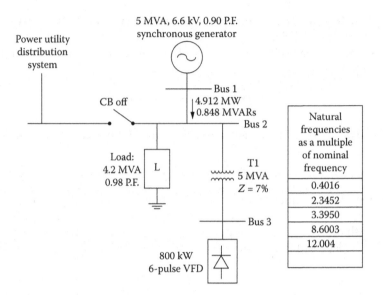

FIGURE 4.3
Turbine-generator set example.

TABLE 4.1

Current and Electrical Torque Spectra for Branch Bus 1 to Bus 2

Harmonic Order	Current Amplitude (%)	Electric Torque Amplitude (%)
1	100	100
5	2.247	
6		1.162
7	1.492	
11	0.89	
12		0.8137
13	0.742	
17	0.558	
18		0.524
19	0.497	
23	0.408	
24		0.390
25	0.375	
29	0.322	
30		0.313

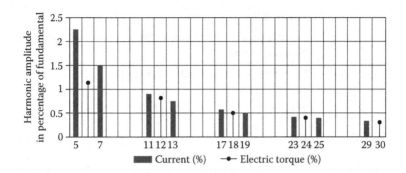

FIGURE 4.4
Harmonic spectrum showing calculated electric torque amplitudes for the described example.

of harmonic pairs 5–7, 11–13, 17–19, etc., which are characteristic of the six-pulse converter and are shown in boldface type in Table 4.1. The elevated torques at these frequencies can be regarded as mechanical resonant spectral components when coinciding with the assumed natural frequencies of the system indicated in Figure 4.3. This situation may lead to a severe increase in vibration amplitudes.

It is important to state that the calculations presented involved a THD at the synchronous generator terminals of around 3.2%, well below the 5% permitted by standard IEEE 519,[7] and the nonlinear load was less than 50% of the rated generator power.

Taking only the eigenvector corresponding to the natural frequency around the 12th-order harmonic, the vibration mode shapes depicted in Figure 4.5 are calculated.

From the shape mode deflections, we can anticipate major oscillations between elements 2 and 3 due to their different polarities. This means that

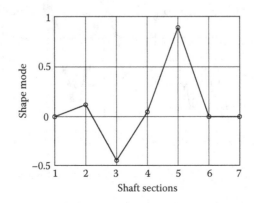

FIGURE 4.5
Calculated vibration mode shapes for the 12th harmonic natural oscillation frequency.

TABLE 4.2

Calculated Torques for the Different
Shaft Sections

Shaft Section	Shaft Torque in p.u.
1–2	0.019
2–3	0.411
3–4	0.145
4–5	00.14
5–6	0.031
6–7	1.4×10^{-5}

torques act in opposite directions. The lowest value of shaft torque can be expected in elements 6 and 7, where there is almost no mode shape deflection between these two elements. Shaft torques will be expected in all other shaft sections. Using Equation (4.15), the shaft torques, T_m, for all the shaft sections are calculated and presented in Table 4.2.

$$T_m = K_{(i,i+1)}(\theta_{i+1} - \theta_i) \tag{4.15}$$

Figure 4.6 shows the calculated shaft torque in the different shaft sections. The mechanical damping is usually low and depends on mechanical design and operative condition. Thus, from the previous results it appears that a possible way to avoid a mechanical resonance could be achieved by modifying the natural frequencies of the mechanical system—for example, by diminishing the inertia (J) and spring (K) constants to reduce residual vibrations

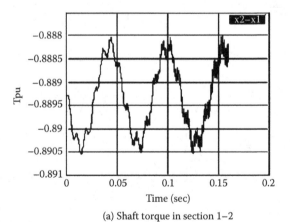

(a) Shaft torque in section 1–2

FIGURE 4.6
Calculated shaft torques for the example of Figure 4.2. *(Continued)*

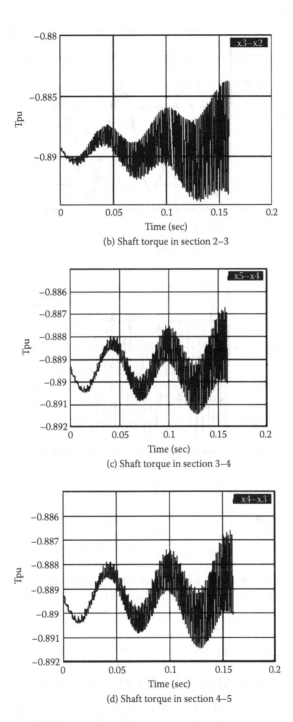

FIGURE 4.6 (Continued)
Calculated shaft torques for the example of Figure 4.2. *(Continued)*

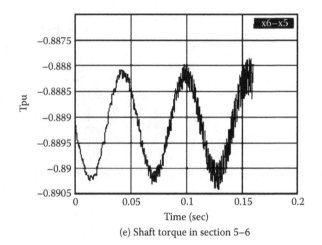

(e) Shaft torque in section 5–6

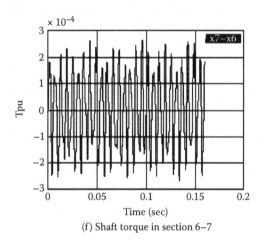

(f) Shaft torque in section 6–7

FIGURE 4.6 (Continued)
Calculated shaft torques for the example of Figure 4.2.

further. This would require a joint and open participation of manufacturers and industry to define practical scenarios and take appropriate actions.

4.9 Abnormal Operation of Solid-State Devices

Under unexpected circumstances, harmonic distortion can also lead to multiple zero crossings of the voltage waveform disturbing the operation of control systems that use the zero crossing as a timing or counter reference.

Additional problems may include interference on motor speed control-lers and abnormal VFD operation when rms voltage or current values are altered by harmonic distortion. This often leads to the need to apply reme-dial measures at the source side of the converters or readjust the protection threshold settings.

4.10 Considerations for Cables and Equipment Operating in Harmonic Environments

In light of a steady increase in harmonic distortion in power systems, the specifications and selection criteria of electrical equipment in industrial installations should be revisited. For example, when distributed generators operate in facilities in which nonlinear loads predominate, the response of conventional generator designs with large subtransient reactances has been shown to be ineffective and often ends in equipment failure after exposure to severe thermal stresses. There must be a threshold for operating param-eters that allows equipment to withstand worst case operating conditions regarding harmonic content and its adverse effects. This goes beyond the theoretical steady-state operation mostly assumed when specifying equip-ment and network components.

4.10.1 Generators

Generators used in the electric power industry are fundamentally designed to feed linear loads. However, when the type of load is predominantly non-linear, generation systems must comply with certain requirements that allow them to operate in stable conditions and without being exposed to excessive heating and torsional torque vibrations, which can make them exceed their permissible operating limits.

Essentially, a nonlinear load produces a voltage waveform distortion at the generator terminals; this imposes the following consequences in the opera-tion of a generator:

- Production of positive- and negative-sequence current contribu-tions that generate torsional torque and vibration mode shapes on the motor shaft. The thermodynamic forces created on the rotor can prematurely wear out shaft bearings.
- Voltage waveform distortion on the supply circuit to the excitation system; this can produce voltage regulation problems.
- Excessive negative-sequence currents; these can contribute to increased voltage unbalance.

As a reference, the following list describes some of the characteristics of synchronous generators found to perform adequately in land and offshore oil well electrosubmersible pump (ESP) applications. This relates to cases where VFDs make the largest portion of the load and where isolated DGs source all or most of the demanded power:

Independent static excitation system is present.

Rotor is furnished with ammortisseur (damper) copper bars.

Transient reactance, X_d', is between 16 and 18%.

Direct axis subtransient reactance, X_d'', is between 13 and 15%. Notice that these values are considerably lower than typical synchronous generators for linear load applications, which are designed with subtransient impedances around 25%.

Insulation (rotor and stator) is according to American Petroleum Standard 546.

Power factor 0.85 is lagging. Here it is important to stress that industrial systems that involve large VFDs or significant numbers of small units may run at power factors close to unity and in some cases even on leading power factors.

Operation with nonlinear loads is satisfactory. This means that a generator must withstand a maximum THD of 8% across its terminals. Notice that this THDV is in excess of the 5% recommended by IEEE for general distribution systems, 120 to 69,000 V, as discussed in Chapter 3.

An open-circuit voltage shows waveform distortion inferior to 2%.

Generators are provided with an oversized damper winding consisting of copper bars to properly handle the additional heating caused by harmonic currents.

Static brushless-type excitation systems with (around 10%) oversized rectifying diodes are present.

Permanent magnet generator (PMG) types of excitation systems are present.

The regulator must be able to handle harmonic distortion typical of 12-pulse types of converters.

This list reveals features of synchronous generator designs that industry should consider to ensure satisfactory performance of generating units in electrical environments highly exposed to harmonics.

4.10.2 Conductors

Power conductors used in distribution systems must be able to carry fundamental and harmonic currents without developing conductor overheating that would be translated to excessive losses. For this, it is important to select conductor sizes considering a permanent steady-state condition over current factor of at least 125%, following National Electrical Code (NEC) 1996, articles 430-24 and 220-10(b). In the latter article, it is recommended that conductor sizes be chosen to withstand 125% of the continuous currents plus the nonpermanent ones.

Also, in installations with shielded cables where shielding is grounded at intervals, it is important to consider a margin to account for the effect of induced currents in the power conductors. An additional 10% to the current specified in the former paragraph holds reasonable. However, oversizing conductors to take additional currents up to 100% of rated values in some VFD applications is sometimes common in industrial networks. This may occur in cases when the electric networks are designed for accommodating future load expansions.

For networks operating in harmonic environments, use of specialized software to determine conductor ampacity in the presence of waveform distortion under worst case scenarios is highly recommended. These results can also be used for protection coordination purposes.

4.10.3 Energy Metering Equipment

The impact that harmonic distortion can have on induction disk meters is an area of present debate in the technical community. The question that technical working groups are trying to resolve is not simply whether the induction disk in a watt-hour meter runs faster or slower, because these conditions may change, depending on the magnitude and order of existing harmonic and DC offset on voltage and current signals at the metering point. Efforts are focused on trying to define what actions can allow the improvement of active and reactive power metering under severe harmonic distortion.

Fortunately, utilities show increasing concern for timely detection of excessive harmonic distortion situations and are seeking compliance with recommended limits through the application of corrective actions. The application of harmonic mitigation measures at specific locations where nonlinear loads exist will actually prevent extreme voltage distortion from penetrating the network and reaching metering points at adjacent facilities.

References

1. *Annual Energy Outlook 2014 with Projections to 2040*, DOE/EIA-0383, April 2014.
2. IEEE 18, IEEE Standard for Shunt Power Capacitors, Institute of Electrical and Electronic Engineers, October 2014.

3. Wagner, V.E., Balda, J.C., Griffitki, D.C., McEachern, A., Barnes, T.M., Hartmann, D.P., Phileggi, D.J., Emannuel, A.E., Horton, W.F., and Reid, W.E., Effects of Harmonics on Equipment, *IEEE Trans. Power Delivery*, 8(2), 1993, report of the IEEE Task Force on the Effects of Harmonics on Equipment.
4. IEEE Interharmonic Task Force, Cigré 36.05/CIRED 2 CC02 Voltage Quality Working Group, *Interharmonics in Power Systems*, December 1997.
5. Pelly, B.R., *Thyristor Phase-Controlled Converters and Cycloconverters*, John Wiley & Sons, New York, 1971.
6. Arrillaga, J., Bradley, D.A., and Bodger, P.S., *Power System Harmonics*, John Wiley & Sons, New York, 1985.
7. IEEE 519-1992, *Recommended Practices and Requirements for Harmonic Control in Electric Power Systems*.
8. Escobar, J.C., and De La Rosa, F., Shaft Torsional Vibrations due to Nonlinear Loads in Low-Capacity Turbine Units, in *2001 IEEE Power Engineering Society Meeting Proceedings*, Vancouver, BC, July 15–19, 2001.

5

Harmonic Measurements

5.1 Introduction

The entire issue of power system measurements is aimed at collecting relevant data for assisting utility planning and operation in a number of aspects key to the efficient transmission and distribution of electric energy. It is also intended to provide reliable energy consumption metering at industrial, commercial, and residential facilities.

A glimpse into the operation of a power system can allow us to realize the many instances when measurements are required. For example, the energy trading among different interconnected utilities/cooperatives requires reliable power delivery measurements that quantify the number of energy blocks that are bought and sold. The substation engineer looks at multiple panel instrumentation to guarantee that voltage and frequency are kept within specified limits, and that the current on the different feeders follows the predicted demand, which must match the capacity of the substation transformer banks. Power factor is also observed to ensure a proper balance between active and reactive power to minimize losses in the distribution system.

As loads fluctuate during the day in response to different demand patterns, utilities switch capacitor banks on and off to keep the voltage profile within tolerable limits. Under light load conditions, there is no need for reactive power compensation; this typically occurs during nighttime. As the load picks up, so does the voltage drop along distribution feeders, and at some distance from the substation, voltage may tend to decrease below permissible limits. It is then when strategically placed capacitor banks or inductive voltage regulators are "switched on" to help raise the voltage profile back to nominal values. The active and reactive power measurements at the substation are therefore key to energy dispatch operators to keep voltage regulation within tight limits.

Another relevant measurement aspect is protection device coordination, which follows preestablished settings that allow protective devices to trip as a response to large currents identified as faults. However, some of them, namely, distance relays, have the capability to carry out current measurements, and using the information from the fault current, they can provide an approximate distance to the fault. Supervisory control and data acquisition (SCADA) systems communicate with substation and feeder remote terminal units, smart relays, and substation automation systems to monitor real-time status of the network and provide remote control of devices such as circuit breakers, capacitor banks, and voltage regulators. The list involving measurement and monitoring of electric parameters can go on and on.

With the steady increase of solid-state electronics in industrial, commercial, and residential facilities, utilities face an increasingly challenging task to carry out reliable measurements due to the waveform distortion on voltage and current signals. This unfair race between fast-growing customers joining the crowd of harmonic generators and utilities struggling to adapt appropriate measuring schemes can soon leave utilities far behind. Decisive efforts to control harmonic currents within industry limits before they converge at and disturb distribution substation monitoring equipment must be undertaken.

At harmonic source locations, the problem can be even worse. The unfiltered higher-frequency components of current at harmonic-producing loads may not give rise only to measurement equipment inaccuracies when they reach or exceed certain thresholds.[1,2] They can produce communication interference,[3] equipment heating problems, false protective device tripping, and even instability conditions on voltage regulation systems in synchronous generators. This is particularly true in installations where customer substation transformers are loaded with mostly nonlinear loads.

An even more delicate problem arises when a customer generates significant harmonic waveform distortion that affects adjacent utility customers. Because all customers can be regarded as harmonic producers to some extent, utilities may find it difficult to pinpoint the right location of the harmonic source, particularly when resonant networks come into play. Utilities may need to carry out measurements at a number of locations involving suspected customers before they can decide the source of the problem to start discussing remedial measures.

Aspects that require a careful standpoint are those related to adequate measurement periods, transducers, and most suitable measuring equipment. This chapter provides a general discussion on the most significant aspects to observe regarding harmonic measurements.

5.2 Relevant Harmonic Measurement Questions

5.2.1 Why Measure Waveform Distortion?

From the electric utility perspective, the general objectives for conducting harmonic measurements may be summarized as follows:

To verify the order and magnitude of harmonic currents at the substation and at remote locations where customer harmonic sources may be affecting neighboring installations

To determine the resultant waveform distortion expressed in the form of spectral analysis

To compare the preceding parameters with recommended limits or planning levels

To assess the possibility of network resonance that may increase harmonic distortion levels, particularly at or near capacitor banks

To gather the necessary information to provide guidance to customers in controlling harmonic levels within acceptable limits

To verify efficacy of implemented harmonic filters or other corrective schemes

To determine tendencies in the voltage and current distortion levels on a daily, weekly, monthly, etc., basis

5.2.2 How to Carry Out Measurements

As mentioned in Chapter 3, IEEE 519-1992[3] provides a general description of key features to take into account when conducting harmonic measurements. It does not, however, underline other aspects regarding duration, integration times, and statistical treatment of collected data. In its current distortion limits section, IEEE 519 suggests that the 15 (or 30) min maximum demand averaged over a 12-month period should be used as the load current, I_L, to determine the ratio, I_{sc}/I_L. This important aspect warns utilities about the need for keeping monthly records of maximum demand to assess total demand distortion properly.

IEC 61000-4-7, Edition 2,[4] considers the measurements of voltage and current to obtain spectral content up to the 40th harmonic using 200 ms measurement windows. The preferred test instrument must be based upon rms calculation of each performance index over a synchronous contiguous 12-cycle window. The 12-cycle window has been adopted in the International Electrotechnical Commission (IEC) standards for 60 Hz systems. These 12-cycle data can then be processed into 3 s, 10 min, and 2 h interval data for each index. Note that rms index values would tend to decrease if larger measurement intervals were used.

Measurement windows are grouped and smoothed using a 1.5 s first-order filter, whose value (for each individual harmonic group) is compared against the limits established in the four test classes (A through D) of IEC 61000-3-6.[5,6] Power measurement is included in the setup because it is the basis for limit calculations for class D equipment. Thus, the measurement equipment is rather sophisticated because it must meet stringent design requirements.

Compatibility levels presented in Chapter 3 refer to harmonic levels sustained for periods up to 3 s to account for interference on sensitive electronic devices and up to 10 min to account for thermal effects on miscellaneous equipment and cables. Therefore, harmonic measurements must consider these needs to seek compliance with compatibility levels.[7]

5.2.3　What Is Important to Measure?

If a utility engineer needed to decide the parameters to consider in evaluating harmonic distortion problems, most likely the decision would involve voltage and current waveforms. This is indeed the right choice because other parameters, such as real, reactive, and total power, energy, and even unbalance, can be calculated from these two quantities. As discussed in Chapter 1, distorted voltage and current waveforms can be expressed as Fourier or other time series. Harmonic distortion and all power quality indices described in Section 1.5 can in fact be determined from these two basic parameters. Nevertheless, power quality monitoring equipment is presently designed to directly provide peak and true rms voltage, current, and power quantities, along with harmonic indices comprising total and individual harmonic distortion and transformer K factor, among others.

Interharmonics, noninteger spectral components, and subharmonics, spectral components with frequencies below the fundamental power frequency as described in Chapter 2, are not easy to characterize. IEC 61000-4-7[4] provides definitions and signal processing recommendations for harmonic and interharmonic measurements.

5.2.4　Where Should Harmonic Measurements Be Conducted?

Harmonic distortion occurrence in an electrical installation can sometimes be assessed through a simple inspection of the types of loads at a given customer installation. All this requires is familiarity with the characteristic harmonic spectrum of each type of common nonlinear load, as described in Chapter 2. However, considering additional waveform distortion caused by transformer saturation or resonant conditions, a more precise evaluation should be carried out. This involves direct measurements at selected locations—for example, the point of common coupling (PCC) described in Chapter 3 and the node where nonlinear loads are connected.

It is understandable that the main location where measurements are to be conducted is the customer-utility interface. This is so because compliance with harmonic limits must be verified at this location. In customer-owned transformer locations, the PCC is the point where the utility will meter the customer, generally the high-voltage side of the transformer. If the utility meters the low-voltage side, then this becomes the PCC.

Also, measurements at LV-connected equipment locations are required when compliance with IEC 61000-3-2[8] (which covers all electrical and electronic equipment with an input current up to 16 A per phase) is sought or when harmonic filtering schemes must be designed at nonlinear loads locations. This is more likely to occur in the industrial or commercial environment where large harmonic-producing loads are operated and served from transformers feeding other sensitive loads. Other instances in which harmonic measurements would be required are when studies are conducted to determine the reasons for abnormal operation or premature failure of equipment, unexpected relay protection tripping, or excessive telephone interference.

5.2.5 How Long Should Measurements Last?

The decision on the optimal period to conduct harmonic measurements may appear somewhat complicated. The reasons for this are diverse. In residential circuits, due to similarity in the types of electronic loads, the expected spectral content may be easily characterized in short-term measurements. However, care must be exercised when the feeder that supplies residential customers is the same from where large commercial/industrial installations are served. If commercial installations are involved, it may be possible to anticipate the types of harmonics because they will typically be linked to fluorescent lighting and power sources from diverse LV electronic equipment.

Industrial installations, however, are a special case because they are usually composed of a mix of loads having a diversity of spectral contents, which may require long-term measurements to characterize harmonic content. This need may become more obvious if cyclic loads exist because measurements to characterize harmonics at the PCC would need to encompass all, or at least the most significant, duty cycles. Long-term measurements may also be required when investigating or trying to resolve the origin of suspicious disturbances affecting a number of customers.

IEEE 519 guidelines do not specify a definite measurement period for capturing harmonic waveform distortion. Under steady-state operation and where no loading variations occur, a few minutes' recording may be sufficient, and averaging over a few seconds should meet the requirements. However, due to the changing nature of loads in most situations, measurements over a few days may be needed to ensure that load variation patterns and their effects on harmonic distortion are considered.

IEC 61000-2-2[9] suggests assessment periods of 1 week for 10 min values regarded as short-time effects (Uh, sh) and of 1 day for 3 s values regarded as very short-time effects (Uh, vs), and to allow taking into account daily work shift patterns and participation of different types of loads in the data collection. Long-term effects relate to thermal effects on different kinds of equipment, such as transformers, motors, capacitor banks, and cables from harmonic levels sustained for at least 10 min. Very short-term effects relate to disturbing effects on vulnerable electronic equipment by events lasting less than 3 s, not including transients. Statistical handling of data is carried out in the form of 95 or 99 percentile of daily or weekly values, per EN 50160-1999[10] and CIGRE C4.07/CIRED.[7] Average values of this parameter are then compared with percent of Uh from IEC 61000-3-6 emission limits.[5]

5.3 Measurement Procedure

5.3.1 Equipment

The process demands that recording instruments as well as voltage and current transducers comply with certain characteristics to ensure that representative samples will be obtained.[3] The analog input bandwidth relates to the frequency limit above which the signal is attenuated by more than 3 dB (29.2%). IEEE 519[3] recommends that the bandwidth of 3 ± 0.5 Hz between the −3 dB points with a minimum attenuation of 40 dB at a frequency of f_h + 15 Hz should be used. A 1.5 kHz analog input bandwidth would limit the harmonic measurement up to the 25th harmonic in a 60 Hz and to the 30th in a 50 Hz system. This covers most frequencies of interest in practical applications. Considering Nyquist criterion, if the input signal contains frequencies higher than half the sampling frequency, the signal cannot be correctly interpreted and an analog input bandwidth greater than 3 kHz will be required.

For all harmonic currents below the 65th (3.9 kHz in a 60 Hz or 3.25 kHz in a 50 Hz system) to be processed properly, the sampling frequency should be at least twice the desired input bandwidth, or 8000 samples per second in this case, to cover 50 and 60 Hz systems. The requirement is for 95% or better accuracy and minimum required attenuation of 50 to 60 dB for 30 Hz, 30 to 50 dB for 120 to 720 Hz, 20 to 40 dB for 720 to 1200 Hz, and 15 to 35 dB for 1200 to 2400 Hz signals. The lower limit is for frequency domain, and the higher limit is for time domain instruments.[3] These limits have to do with the attenuation of high-frequency signals when the instrument is tuned at the fundamental frequency.

A large variety of instrumentation exists that can be used to carry out measurements and long-term recordings. Power quality analyzers are capable of carrying out measurements of rms voltage and current and perform calculations of active, reactive, and apparent power. They also compute harmonic distortion of voltage and current signals presenting individual and total harmonic levels, and some of them can calculate $V * t$ and $I * t$ products and K factor. There are indoor and outdoor versions of monitoring equipment, and some of them can be set up to carry out long-term recordings.

5.3.2 Transducers

These elements convert the parameter to measure in a signal of adequate amplitude to be processed by the measuring equipment. However, not only amplitude is important. It is essential that their frequency response have an appropriate bandwidth so as not to produce any signal distortion. As transducers that need to comply with these requirements, the following can be mentioned:

Potential transformers (PTs)

Current transformers (CTs)

Depending on the system voltage and the network configuration and type of load, the voltage can be measured directly or through the PTs. With regard to current measurements, they can be carried out on the primary side using the current probes furnished with the measuring equipment or at the low-voltage side, usually at the utility meter location. Under uncertainty regarding their frequency response, transducers should be subjected to tests to determine that their bandwidth is adequate to carry out harmonic measurements.

Although IEEE 519[3] points out that most utility measuring PTs can be used with a precision of 97% in the frequency range up to around 5 kHz, it is recommended that tests be conducted on TPs to determine that their bandwidth is appropriate up to the frequency of interest. In the case of CTs (those installed at the substation by the power utility for electric current and watt-hour measurements), they have a frequency bandwidth up to 20 kHz with an error smaller than 3%, according to reference 3. Properly grounded (complying with IEEE 518-1982) shielded coaxial cables are recommended for short distances to the measurement equipment. If distances are large over a few tens of meters, fiber optic links are highly recommended to avoid all types of interference on the sometimes small-amplitude signals.

5.4 Relevant Aspects

Harmonics and lamp flicker increase with the use of power electronic devices in the system and create problems for loads susceptible to power quality problems. Standards for power quality and measurements to locate the source of problems are needed to maintain power quality.

The observed practice in relation to the monitoring of electrical parameters shows that measurements should be carried out in at least the PCC and at nodes where nonlinear loads are connected. The former location is important because compliance with standards must be sought at the PCC, and the latter is significant to verify emission limits right at nonlinear load locations. This can be done actually at network nodes that group a number of similar nonlinear loads or at individual points with a single, large-size harmonic-producing load.

We should keep in mind that if harmonic filtering is considered an option in a large industrial facility, cost will often determine the location of filters and thus the point to monitor. This allows verification of the efficacy of the filtering scheme applied.

If interharmonics (noninteger multiples of fundamental frequency found in cycloconverters, arc welders, and electric furnace applications) are of interest, power quality monitoring equipment with the adequate bandwidth and accuracy must be used.

Harmonic measurements at specific customer sites may provide valuable information to determine compliance of end users with standards. It should be borne in mind, though, that the proliferation of electronic switching practically makes every customer a contributor to the harmonic distortion problem. This only makes it somewhat more complicated for utilities to determine specific responsibilities in the case of noncompliance to limits. Exhaustive measurement campaigns at suspect customer sites involving simultaneous observation of parameters at several sites may be required. Determining the right time for and duration of such endeavors must be conducted following the recommended guidelines.

Multiple sources of harmonic distortion thus require that steps toward characterizing emission levels at specific areas of the network be undertaken by utilities and industry as an effort to obtain an overall picture of potential trouble areas. This need will become increasingly evident as the networks expand and conducting surveys at specific locations turns progressively burdensome. Statistical prediction methods and analysis may need to be used as a helpful tool in this process.

If subharmonics are of interest, proper equipment should be used. Interharmonic limits should be limited below harmonic components. Measuring subharmonics is a challenging task.[12] The reason for this is that the frequency range up to 100 Hz remains very sensitive to spectral leakage problems caused by small synchronization errors; measuring subharmonics should be undertaken with this in mind.

Lastly, as noted in the introductory discussion in Chapter 4, it is important to bear in mind the possibility that harmonic sources may increase on the horizon, for which it becomes important to consider future expansions during harmonic assessment and in the specification of harmonic mitigation equipment.

References

1. Arseneau, R., The Performance of Demand Meters under Varying Load Conditions, *IEEE Trans. Power Delivery*, 4, 1993.
2 Arseneau, R., and Filipski, P., Application of a Three-Phase Nonsinusoidal Calibration System for Testing Energy and Demand Meters under Simulated Field Conditions, *IEEE Trans. Power Delivery*, 3(3), 874–879, 1998.

3. ANSI/IEEE 519-1992, *IEEE Recommended Practices and Requirements for Harmonic Control in Electrical Power Systems.*

4. IEC 61000-4-7, *Electromagnetic Compatibility (EMC)—Part 4-7: Testing and Measurement Techniques—General Guide on Harmonics and Interharmonics Measurements and Instrumentation, for Power Supply Systems and Equipment Connected Thereto*, Edition 2, 2002.

5. IEC 61000-3-6, *Assessment of Emission Limits for Distorting Loads in MV and HV Power Systems*, technical report type 3, 1996.

6. van den Bergh, M., Harmonics and Flicker Requirements and Instrumentation, *Conformity*, August 2004, Input 22.

7. Joint WG CIGRE C4.07/CIRED, *Power Quality Indices and Objectives*, final WG report, January 2004, rev. March 2004.

8. IEC 61000-3-2, *Electromagnetic Compatibility (EMC)—Part 3-2: Limits—Limits for Harmonic Current Emissions (Equipment Input Current ≤ 16 A per Phase)*, 2001–10.

9. IEC 61000-2-2, *Electromagnetic Compatibility—Part 2-2, Environment Compatibility Levels for Low-Frequency Conducted Disturbances and Signaling in Public and Low-Voltage Power Supply Systems*, 2002.

10. CENELEC EN 50160-1999, *Voltage Characteristics of Electricity Supplied by Public Distribution Systems*, European standard (supersedes 1994 ed.).

11. IEEE 518-1982, *IEEE Guide for the Installation of Electrical Equipment to Minimize Electrical Noise Inputs to Controllers from External Sources.*

12. Testa, A., and Langella, R., Power System Subharmonics, in *Proceedings of 2005 IEEE Power Engineering Society General Meeting*, San Francisco, June 12–16, 2005.

6

Harmonic Filtering Techniques

6.1 Introduction

In a general context, we can refer to harmonic filters as passive and active filters. Their essential difference, as illustrated in this chapter, stands on whether they provide a (passive) filtering action within a selected bandwidth or as a result of a real-time (active) monitoring process that leads to the injection of real-time canceling harmonic currents.

One of the most common methods for control of harmonic distortion in industry is the use of passive filtering techniques that make use of single-tuned or band-pass filters. Passive harmonic filters can be designed as single-tuned elements that provide a low-impedance path to harmonic currents at a punctual frequency or as band-pass devices that can filter harmonics over a certain frequency bandwidth.

The more sophisticated active filtering concepts operate in a wide frequency range, adjusting their operation to the resultant harmonic spectrum. In this way, they are designed to inject harmonic currents to counterbalance existing harmonic components as they show up in the distribution system. Active filters comprise DC, AC, series, and parallel configurations. Hybrid filters are a combination of passive and active filtering schemes. Active filtering is so extensive and specialized that it is not possible to entirely cover it within the scope of this book.

This chapter presents a straightforward methodology to design a passive filter based on the relationship between fundamental parameters. It also makes use of the Institute of Electrical and Electronics Engineers (IEEE) guidelines[1,2] for the selection of the filter components, presenting some application examples, some of which use commercial software for harmonic filter analysis.

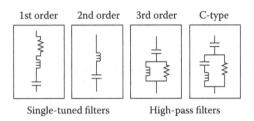

FIGURE 6.1
Electric diagrams of passive filters.

6.2 General Aspects in the Design of Passive Harmonic Filters

Passive filters are the most commonly used filters in industry. As illustrated in Figure 6.1, the following can be found under this category:

Single-tuned filters

High- (or band-) pass filters (first, second, and third order)

6.3 Single-Tuned Filters

Probably the most common harmonic filter in industrial applications, the passive filter, presents very low impedance at the tuning frequency, through which all current of that particular frequency will be diverted. Thus, passive filter design must take into account expected growth in harmonic current sources or load reconfiguration because it can otherwise be exposed to overloading, which can rapidly develop into extreme overheating and thermal breakdown. The design of a passive filter requires a precise knowledge of the harmonic-producing load and of the power system. A great deal of simulation work is often required to test its performance under varying load conditions or changes in the topology of the network.

Because passive filters always provide reactive compensation to a degree dictated by the voltampere size and voltage of the capacitor bank used, they can in fact be designed for the double purpose of providing the filtering action and compensating power factor to the desired level. If more than one filter is used—for example, sets of 5th and 7th or 11th and 13th branches—it will be important to remember that every individual filter will provide a certain amount of reactive compensation.

As discussed earlier, this filter is a series combination of an inductor and a capacitor. In reality, in the absence of an intentionally integrated resistor, there will always be a series resistance, which is the intrinsic resistance of the series reactor sometimes used as a means to avoid filter overheating. All harmonic currents whose frequency coincides with that of the tuned filter will find a low-impedance path through the filter.

The resonant frequency of this filter can be expressed by the following expression:

$$f_0 = \frac{1}{2\pi\sqrt{LC}} \qquad (6.1)$$

where f_0 is resonant frequency in hertz, L is filter inductance in henrys, and C is filter capacitance in farads.

As later described in Section 6.3.3, the quality factor, Q_f, of the filter is the ratio between the inductive or capacitive reactance *under resonance* and the resistance. Typical values of Q_f fluctuate between 15 and 80 for filters that are used in industry. The following relation is used to calculate the quality factor:

$$Q_f = \frac{X_L}{R} = \frac{X_C}{R} \qquad (6.2)$$

Low-voltage filters (480 to 600 V) use iron cores with air gaps that have elevated losses but are associated to low Q_f values. Medium-voltage filters (4.16 to 13.8 kV) have Q_f values in the upper range.

The X/R ratio of low-voltage systems ranges between 3 and 7. These systems do not present an elevated parallel resonant peak in the Z–f characteristic. Although low-voltage filters have elevated losses, they also provide greater attenuation to any oscillation present in the system.

The process of designing a filter is a compromise among several factors: low maintenance, economy, and reliability. The design of the simplest filter that does the desired job is what will be sought in the majority of cases.

The steps to set up a harmonic filter using basic relationships to allow for a reliable operation can be summarized as follows:

- Calculate the value of the capacitance needed to improve the power factor and to eliminate any penalty by the electric power company. Power factor compensation is generally applied to raise the power factor to around 0.95 or higher.
- Choose a reactor to tune the series capacitor to the desired harmonic frequency. For example, in a six-pulse converter, this would start at the fifth harmonic and it would involve lower frequencies in an arc furnace application, as illustrated by the harmonic spectrum of Figure 2.19.
- Calculate the peak voltage at the capacitor terminals and the rms reactor current.

- Choose standard components for the filter and verify filter performance to assure that capacitor components will operate within IEEE 18[2] recommended limits. This may require a number of iterations until desired reduction of harmonic levels is achieved.

Passive filters carry a current that can be expressed as a fraction of the load current at fundamental frequency. As for their cost, they are more expensive than series reactors often used to provide some harmonic attenuation, but they have the advantage of providing reactive power at fundamental frequency. For practical purposes, they are substantially used in industry.

Filter designs usually offer a robust mechanism that provides some minor filtering action for a fraction of other harmonic currents whose order is close to the tuning frequency, provided that no filters tuned at those frequencies exist.

Filter impedance must be smaller than that presented by the system at the tuning frequency. In low-voltage systems in which ratio X/R is small, an individual filter may be sufficient to provide the necessary attenuation. For example, neglecting the intrinsic resistance of the series reactor in a harmonic filter, the lowest value of the impedance–frequency characteristic in Figure 6.2, as seen from the source, results from the resistive component of the system. The location of this point on the y-axis at the tuning frequency would be around three times higher for a network with an X/R ratio of 3 as compared with a case in which X/R is equal to 10. A resistive component with a theoretical zero resistance would make the filter absorb the entire harmonic current of frequency equal to the tuning frequency of the filter. Sometimes a series resistive component is included to control the maximum current allowed through the filter. This will have an impact on the quality factor of the filter, as described by Equation (6.2).

The study of the response of single-tuned filters reveals the following relevant aspects:

Single-tuned filters act as a small impedance path, effectively absorbing the harmonic currents for which they are tuned. It is important to be aware that these filters may take currents of neighboring

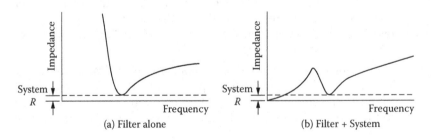

(a) Filter alone   (b) Filter + System

FIGURE 6.2
Frequency response of a single-tuned harmonic filter.

frequencies, as will be shown in this chapter. Therefore, it will be important to assess the impact of those additional currents on the operational parameters of the filter.

For a typical power utility or industrial system, there is a pronounced increase in the impedance slightly below the series LC filter tuning frequency, as shown in Figure 6.2(b). This is a result of a parallel resonant condition between the capacitor of the filter and the source side inductance.

6.3.1 Design Equations for the Single-Tuned Filter

The impedance of the filter branch is given by

$$Z = R + j\left[\omega L - \frac{1}{\omega C}\right] \tag{6.3}$$

where R, L, and C are the resistance, inductance, and capacitance of the series-connected filter elements, respectively, and ω is the angular frequency of the power system.

The series resonance condition is excited when the imaginary part of the impedance is equal to zero, where the only impedance component left is the resistance. The frequency at which the filter is tuned is then defined by the value of ω that makes inductive and capacitive reactance cancel one another in Equation (6.3). This frequency is given by Equation (6.1). If we make h the ratio between the harmonic and the fundamental frequencies of the system, the inductive and capacitive reactances at the harmonic frequency can be expressed as

$$X_{Lh} = h\omega_L \tag{6.4}$$

$$X_{Ch} = \frac{1}{(h\omega_C)} \tag{6.5}$$

Expressed in a different way, assuming zero resistance, the condition for the impedance in Equation (6.3) dropping to zero at the tuning frequency requires

$$X_{Lh} = X_{Ch} \tag{6.6}$$

Substituting Equations (6.4) and (6.5) in Equation (6.6) and solving for h, we get

$$h^2 = \frac{X_C}{X_L}$$

or

$$h = \sqrt{\frac{X_C}{X_L}} \qquad (6.7)$$

6.3.2 Parallel Resonant Points

As mentioned earlier, the interaction of the filter with the source imped-ance (*Ls*) always results in a parallel resonance characterized by the large-impedance peak illustrated in Figure 6.3. Seen from the capacitor bank

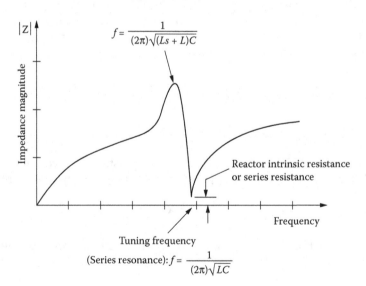

FIGURE 6.3
Resonant points on a single-tuned series RLC filter.

upstream (toward the source), a parallel resonance is to be established at a frequency

$$f\_par\_res = \frac{1}{2\pi\sqrt{(Ls+L)C}} \tag{6.8}$$

which falls slightly below the filter tuning frequency, as illustrated in Figure 6.3.

In installations in which multiple single-tuned filters are required, a parallel resonant frequency will exist for every individual passive filter. Notice that the frequency of the parallel resonant point will experience a shift whenever changes in filter elements L or C or in source inductance Ls occur. Ls can change, for instance, following the disconnection or addition of a transformer at the substation. This could take place every time the power utility changed the configuration of source during transformer maintenance actions or whenever transformers are added to the bank.

A change that can also affect the parallel resonant frequency in Equation (6.8) is the addition of power factor capacitor units on the feeder that serves the nonlinear load where the harmonic filter is installed.

However, the most notable impact of the source impedance on the filter performance is its parallel impedance peak. Figure 6.4 illustrates the

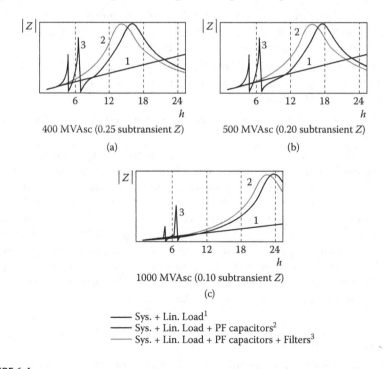

400 MVAsc (0.25 subtransient Z)

(a)

500 MVAsc (0.20 subtransient Z)

(b)

1000 MVAsc (0.10 subtransient Z)

(c)

—— Sys. + Lin. Load[1]
—— Sys. + Lin. Load + PF capacitors[2]
—— Sys. + Lin. Load + PF capacitors + Filters[3]

FIGURE 6.4
Response of a harmonic filter for different source MVAsc.

impedance value at parallel resonant peaks in a two-branch filter design as a function of source impedance. This is shown in three different plots with increasing short-circuit megavoltampere values (or reduced source impedance) in Figure 6.4. Observe how the parallel impedance peaks are reduced in amplitude, which will have the effect of decreasing harmonic distortion produced by any harmonic current component of frequency smaller than the tuning frequency of the filter. The opposite effect would be observed when the small source impedance in a parallel transformer configuration on the source side increases as one of the transformers is taken out for maintenance.

A problem that can arise with the adjacent parallel resonant points is a detuning action. If the filter is tuned at exactly the frequency of interest, then a shifting of the series resonant point to higher-frequency values will result in a sharp impedance increase, as seen by the harmonic current of that frequency order. This can occur, for example, from capacitor aging, which would cause some decrease in capacitance. If the parallel-resonance peak shifts in such a way that it aligns with the frequency of a characteristic harmonic of the load, the resultant harmonic voltage amplification can be disastrous because it can produce overvoltage stresses on solid insulation of cables and on machine windings. The aspects involved in the detuning action of the filter can be described as follows:

The tripping action of capacitor bank fuses disconnecting one or two single-phase units will decrease the equivalent three-phase capacitance, increasing the tuning frequency of the filter.

Manufacturing tolerances of the filter elements can result in a shifting of the tuning frequency in any direction, for which it is important to take them into account.

Temperature variations can produce an accelerated aging on the capacitor units.

The variations on the topology or configuration of the system, which changes the upstream inductive reactance seen from the location of the filter, can also have an impact on the location of parallel resonant points.

Considering the preceding points and assuming a fixed source impedance, it is convenient to tune the filter at a frequency slightly below the desired frequency, typically 3 to 5%. This will account for small tune frequency shifts to higher values over time, yet allow the filter to provide a low-impedance path. Also, an unbalance detection scheme to protect the capacitor bank and to ensure the proper operation of the filter will be important to consider.

6.3.3 Quality Factor

Regarding single-tuned harmonic filters, the quality factor relates the ability of a filter to dissipate the absorbed energy at the tuned frequency. IEEE[1]

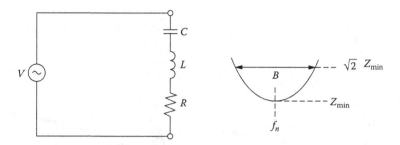

FIGURE 6.5
Fundamental quantities that determine the performance of a passive single-tuned filter.

quotes an approximate equivalent definition as the ratio of the resonant frequency, ω_θ, to the bandwidth between those frequencies on opposite sides of the resonant frequency where the response of the resonant structure differs 3 dB from that at resonance:

$$Q = \frac{\omega_\theta}{(\omega_1 - \omega_2)} \tag{6.9}$$

In a resistive–inductive–capacitive (RLC) series circuit, we can define Q as

$$Q = \frac{1}{R}\sqrt{\frac{L}{C}} = \frac{X_{Lh}}{R} = \frac{X_{Ch}}{R} \tag{6.10}$$

where X_{Lh} and X_{ch} are the inductive and capacitive reactances, respectively, at the resonant frequency of the series filter. Figure 6.5 and Table 6.1 summarize the basic parameters that describe the single-tuned passive filter.[3]

Figure 6.6 shows a number of plots for a harmonic filter with different Q_f values. The shaded area delineates the response of the filter. Notice how the larger the Q_f, the better the filtering action achieved, which is reflected on the lowest impedance at the tuning frequency. This is a logical effect that results

TABLE 6.1

Relevant Quantities on a Passive Single-Tuned Filter

Tuned Harmonic Order	Quality Factor	Bandwidth	Reactive Power at f_1	Active Power at f_1 (losses)
$h = \dfrac{f_n}{f_1} = \sqrt{\dfrac{X_C}{X_L}}$	$Q_f = \dfrac{n \cdot X_L}{R} = \dfrac{X_C}{n \cdot R}$	$B = \dfrac{f_n}{Q_f}$	$Q_C = \dfrac{V^2}{X_C} \cdot \dfrac{n^2}{(n^2 - 1)}$	$P \cong Q_C \cdot \dfrac{n}{n^2 - 1} \cdot \dfrac{1}{Q_f}$

Note: f_1 = fundamental frequency, $\omega = 2\pi f_1$ = angular frequency, f_n = tuning frequency, n = harmonic order f_n/f_1, V = nominal line-to-line voltage, X_L = inductor reactance at fundamental frequency = $L\omega$, and X_C = capacitor reactance at fundamental frequency = $1/\omega C$.

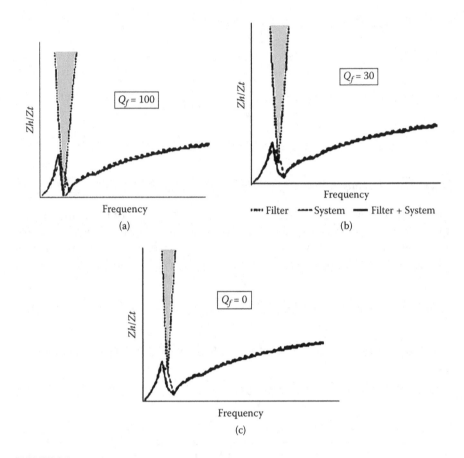

FIGURE 6.6
Impedance-frequency plots for a harmonic filter for different values of Q.

from decreasing the purely resistive impedance at the resonant frequency in Equation (6.10).

The following points summarize the most relevant quality factor aspects in single-tuned filters:

Typically, the resistance of a single-tuned harmonic filter is the intrinsic resistance of the reactor.

However, R can be favorably used to vary the quality factor of the filter and provide a way to control the amount of desired harmonic current through it.

A large Q_f value implies a prominent valley at the resonant (tuning) frequency of a filter, and therefore the trapping of the largest amount of harmonic frequency.

The best reduction of harmonic distortion will be achieved with large Q_f value filters. However, care should be exercised in assessing harmonic currents of frequencies other than the one for which the filter is tuned because they will also find a reduced impedance path. These currents will provide increased heat dissipation. It will often be necessary to conduct computer-aided harmonic simulation studies to predict the performance of the filters, especially when multiple harmonic sources exist.

Lower-quality factor filters could be used in situations in which harmonic distortion barely exceeds the limits and a small filtering action is all that is needed to bring performance into compliance.

6.3.4 Recommended Operation Values for Filter Components

6.3.4.1 Capacitors

Capacitor banks are voltage-sensitive components of filters for the following reasons:

Capacitors may be exposed to increased voltage during waveform distortion produced by harmonic components.

The voltage increase can be in the form of an augmented peak or an increase of the rms value.

The reactive power output of the capacitor will change with the square of the rms-distorted to the rms-undistorted voltage ratio.

In situations in which switching overvoltages are frequent, filter capacitors can be exposed to damage.

IEEE 18[2] recommends limits for the operation of shunt capacitors in power systems, including current, power, and voltage across the capacitor units. Following this guideline, capacitors are designed to be operated at or below their rated voltage and to be capable of continuous operation under contingency and normal conditions, provided that none of the limitations of Table 6.2 are exceeded.

TABLE 6.2

Maximum Recommended Limits for Continuous Operation of Shunt Capacitors under Contingency Conditions

VAR	135%
rms voltage	110%
Rated voltage, including harmonics	120%
rms current	135%

Note that the limit for the rms current is lower than the 180% that was considered in the 1992 revision of IEEE 18 because that current level may be causing a rated power exceeding the recommended limit. It is important to observe compliance with these limits mainly at facilities where considerable harmonic waveform distortion exists or capacitors are part of a harmonic filter.

Generally, the capacitor voltage in a bank used in a single-tuned harmonic filter will be exposed to an increased voltage that can be approximated in terms of the harmonic order (h) of the filter as follows:

$$V_{cap} = \frac{h^2}{(h^2 - 1)}(V_{system})$$ (6.11)

When the maximum amplification of the voltage is verified, the worst conditions that include the maximum voltage considering the tolerance of the filter elements (typically 8% for the capacitor and 5% for the inductor) must be tested.

When a capacitor bank of a nominal voltage different from that of the system is used, the effective bank, kVAR, must be determined from the following expression:

$$kVAR_{effective} = \left(\frac{V_{systemL-L}}{V_{capL-L}}\right)^2 (kVA_{rated})$$ (6.12)

The presence of a reactor in the filter changes the effective kVAR of the filter. The new output will be

$$kVAR_{filter} = \frac{V_{capL-L}}{(Z_C - Z_L)}$$ (6.13)

If reactive compensation is needed, the capacitor bank of the filter can be chosen to provide it. However, the designer will typically need to follow an iterative process to decide the suitable VARs of the bank.

6.3.4.2 Tuning Reactor

The maximum voltage elevation across the reactor must also be determined. The parameters usually included in the specification of the reactor are the following:

50/60 Hz current

Harmonic current spectrum

Short-circuit current

X/R ratio

System voltage

Basic insulation level (BIL)

Reactors used in harmonic filters are sometimes designed with an air core. This provides linear characteristics regarding frequency and current. In applications that involve industrial power systems, 5% in tolerance is typically used. The relation, X/R, at system power frequency, which is typically smaller than 150, can be further manipulated to obtain the desired quality factor as described previously.

Also, the maximum voltage elevation across the reactor must be determined. The nominal voltage of the reactor must be able to handle the overvoltage imposed under a short-circuit condition, for instance, when a capacitor fails. The basic insulation level (BIL) of the reactor and, similarly, of the capacitor bank must be the same as that of the power transformer that feeds the load where the filters are integrated.

6.3.5 Unbalance Detection

The purpose of an unbalance detection scheme is to remove capacitor banks as soon as phase overcurrent protection trips due to a single fault-to-ground event. Generally, unbalance detection triggers an alarm when one or several capacitor stages are lost in the bank. In a harmonic filter, the failure of a capacitor unit can detune the filter and produce harmonic voltage amplification following a shift in the parallel resonant point.

6.3.6 Filter Selection and Performance Assessment

First, determine if reactive compensation is required. If this is the case, the capacitor bank of the filter must be sized to provide the needed VARs.

For certain system conditions, more than one filter may be needed. Consider all possible scenarios to determine the worst case condition. In certain applications, nonlinear loads may be cyclic and filter schemes must be designed so that they allow the possibility of having filter branch components in and out, as needed.

When analyzing the effectiveness of a filter (understood as the degree of harmonic suppression), it is important to try it for different upstream impedance conditions, which are determined by the utility source impedance.

Another important point to consider is the possibility of shifting the parallel resonant points. As described previously, changing feeder capacitance as a response to load variations or voltage profile can have an impact on parallel resonant frequency. For instance, if at 100% feeder loading there is a system resonant peak at the 4.5th harmonic, a loading decrease accompanied by disconnection of power factor capacitor banks on the same harmonic filters bus may shift the parallel resonant peak to the 5th harmonic. Any fifth harmonic current from the load would see a large upstream impedance, and a higher voltage harmonic distortion at that frequency would develop. This situation must be considered in the design of harmonic filters or in the implementation of special operation rules to minimize the negative effects of the resonant peak.

The traditional design criteria in relation with the presence of harmonic currents in industrial networks are total harmonic distortion (THD) and telephone interference factor (TIF) levels. Harmonic distortion is likely to be exceeded in industrial and commercial applications involving large power converters, massive amounts of fluorescent lighting, and significant amounts of office equipment. Therefore, commercial and financial facilities, office or commercial buildings, and corporate and public offices are good candidates to exceed compliance with recommended limits. Likewise, the telephone interference factor may reach considerable levels in extensive power cable networks under resonance conditions.

THD should be evaluated at every relevant bus in the system—namely, at the main plant substation bus, at those nodes with harmonic current sources, and wherever sensitive equipment exists. If THD limits are above limits, then the need to provide harmonic filters must be sought and THD/TIF levels reassessed.

Operation scenarios to be considered in the filter design stage should include network and load reconfiguration that involve plant expansion and future load growth.

If the rated values of the filter components are exceeded under normal operation, an adjusting action should follow. However, any adjustment performed to the filter scheme should be anticipated and considered in the filter design specifications.

The evaluation of harmonic filters must include power frequency and harmonic losses. This is particularly relevant for the design of a minimum filter, i.e., that which is specified and installed to bring harmonic current distortion within limits but not for power correction purposes.

6.4 Band-Pass Filters

Band-pass filters, high-pass in particular, are known by their small impedance value above the corner frequency. In this type of filters L and R are connected in parallel instead of series. This results in a wide-band filter whose impedance at high frequencies is limited by R. The quality factor of this type of filter is determined by the ratio of R/X_L. The typical frequency response of a high-band-pass filter is shown in Figure 6.7. This filter draws a considerable percentage of frequency harmonic currents above the corner frequency. Therefore, this frequency must be selected below all harmonic currents that have an important presence in the installation.

In planning to adopt a high-pass filter as a harmonic mitigating measure, the following aspects should be considered:

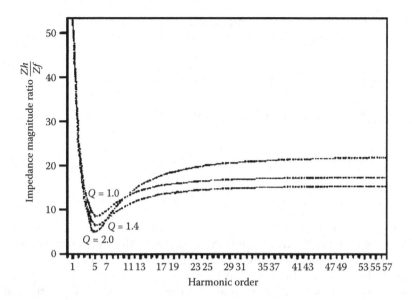

FIGURE 6.7
Response of a high-pass filter for different Q_f values.

The impedance–frequency characteristic of a high-pass filter will entail a very different filtering action than that provided by a single-tuned filter.

Harmonic current elimination using a high-pass filter may require a quite different sizing of filter elements, particularly of the capacitor bank, compared with a single-tuned filter. For example, a 3 MVAR bank used in a fifth harmonic filter in a 60 Hz application may fall short in size when used as part of a high-pass filter with a corner frequency of 300 Hz. Obviously, this will very much depend on the additional harmonic currents that the high-pass filter will be draining off.

Band-pass filters are usually tuned to filter high-order harmonics and can cover a broad range of frequencies. A special case is the C-type high-pass filter (see Figure 6.1), where L is replaced with a series LC circuit tuned at the power system frequency. Its quality factor continues being R/X_L. In this filter, the resistance is evaded at the fundamental frequency, current flows through the resonant LC circuit, and this results in a lossless filter.

The impedance of a band-pass filter can be expressed as

$$Z = \frac{1}{j\omega C} + \frac{1}{\left(\dfrac{1}{R} + \dfrac{1}{j\omega L} \right)} \tag{6.14}$$

The corner frequency of the filter is given by

$$f = \frac{1}{2\pi\sqrt{LC}}$$ (6.15)

The quality factor of the filter is calculated as

$$Q_f = \frac{R}{\sqrt{\dfrac{L}{C}}} = \frac{R}{X_L} = \frac{R}{X_C}$$ (6.16)

For typical high-pass filters, Q_f values between 0.5 and 2.0 are found. Filters with higher Q_f would provide a superior filtering action at the corner frequency, although at higher frequencies, the impedance would show a steady increase, as illustrated in Figure 6.7. Filters with smaller Q_f values would show an inferior performance at the corner frequency, although at frequencies higher than that, a less pronounced increase of impedance with frequency is obtained. This is also illustrated in Figure 6.7.

Other factors that must be considered in the selection of Q_f are the following:

The tuning frequency of the filter

Concerns for telephone interference (if it exists)

Power losses

6.5 Relevant Aspects to Consider in the Design of Passive Filters

A number of aspects must be considered in the design stage of passive filters for controlling problems associated with harmonics. These are summarized as follows:

The capacitive kVAR requirements for power factor correction. Some installations may benefit from the installation of harmonic filters because power factor will be improved. In other situations, power factor correction needs may dictate the size of the capacitor bank to use.

In single-tuned filters, watch the resonant parallel peaks resulting from the interaction between the filter and the source.

Consider tolerances of filter components. They may produce undesirable shifts of resonance frequencies.

Also look for load and network impedance changes that may modify established worst case harmonic scenarios.

Oversized capacitor banks may be required in high-pass filters with low corner frequencies and significant higher-order characteristic harmonics.

Be aware of quality factor filters as a measure to control the amount of harmonic currents to be drawn from the system. Avoid overloading capacitor banks using a series resistor in single-tuned filters. A trade-off between decreased THD values and power factor correction assuring capacitor bank integrity will often decide the Q_f value to adopt in a filter.

Extensive electric networks may have nonlinear loads with different spectral contents. Whenever possible, grouping loads by type of harmonic spectrum (for instance, 6-pulse converters, 12-pulse converters, arcing type devices, fluorescent lighting, etc.) can optimize the installation, location, and sizing of harmonic filters. Although this is a difficult task to achieve, especially when comparable types of loads are not on the same location, the idea should be considered as a way to reduce the number of harmonic filters to install. Load grouping could also help reduce telephone interference by trying to keep telephone lines as distant as possible from sites carrying higher-order harmonic currents.

Minimum filters may be adopted under no reactive compensation needs. The parameters of a minimum filter must be chosen to reach the maximum recommended THD limit.

Always watch for filter power losses.

6.6 Methodology for Design of Tuned Harmonic Filters

The recommended procedure for the design and validation of single-tuned harmonic filters is summarized in the following sections.

6.6.1 Select Capacitor Bank Needed to Improve the Power Factor from the Present Level Typically to around 0.9 to 0.95

The capacitive reactance needed to compensate the needed VARs to improve the power factor from PF_1 (associated with θ_1) to PF_2 (associated with θ_2) is given by

$$\text{VARs} = P (\tan \theta_1 - \tan \theta_2) \tag{6.17}$$

with

$$P = (V)\,(I)\cos\theta_2 \tag{6.18}$$

Sometimes P must be calculated from multiplying the apparent power, S, by the power factor of the load.

The capacitive reactance required is obtained with the following relation:

$$X_{C1} = \frac{V^2}{\text{VARs}} \tag{6.19}$$

where V and VARs are capacitor-rated values.

At harmonic frequency h, this reactance is

$$X_{Ch} = \left(\frac{1}{h}\right) X_{C1} \tag{6.20}$$

6.6.2 Choose a Reactor That, in Series with a Capacitor, Tunes Filter to Desired Harmonic Frequency

The inductive reactance required at harmonic h is, in this case,

$$X_{Lh} = X_{ch} \tag{6.21}$$

and at fundamental frequency, the required reactance is

$$X_{L1} = \left(\frac{1}{h}\right) X_{Lh} \tag{6.22}$$

6.6.3 Determine Whether Capacitor Operating Parameters Fall within IEEE 18[2] Maximum Recommended Limits

6.6.3.1 Capacitor Voltage

The rms and peak voltage of the capacitor must not exceed 110 and 120%, respectively, of the rated voltage. They can be determined as follows:

$$V_{C_{rms}} = \sqrt{(V_{C1}^2 + V_{Ch}^2)} \tag{6.23}$$

$$V_{C_{peak}} = \sqrt{2}(V_{C1} + V_{Ch}) \tag{6.24}$$

where voltage through the capacitor at fundamental frequency is given by

$$V_{C1} = X_{C1} I_{C1} \tag{6.25}$$

I_{C1} is the current through the capacitor, and it is calculated in terms of the maximum phase-to-neutral voltage, which in turn is specified 5% above the rated value, to account for voltage regulation practices:

$$I_{C1} = (1.05)\frac{V_{L-N}}{(X_{C1} - X_{L1})} = (1.05)\left[\frac{\left(\frac{V_{L-L}}{\sqrt{3}}\right)}{(X_{C1} - X_{L1})}\right] \tag{6.26}$$

V_{ch} is found in terms of I_{ch}, which must be determined from measurements or from a typical harmonic spectrum of the corresponding nonlinear load:

$$V_{ch} = X_{ch}I_{ch} \tag{6.27}$$

6.6.3.2 Current through the Capacitor Bank

The rms current through the capacitor bank must be within 135% of the rated capacitor current, to comply with IEEE 18. Its value is determined from the fundamental current and from the harmonic currents under consideration:

$$I_{Crms} = \sqrt{(I_{C1}^2 + I_{Ch}^2)} \tag{6.28}$$

6.6.3.3 Determine the Capacitor Bank Duty and Verify That It Is within Recommended IEEE 18 Limits

$$kVAR = \frac{(V_{Crms})(I_{Crms})}{1000} \tag{6.29}$$

where V_{Crms} is the voltage through the capacitor calculated in Equation (6.23), and I_{Crms} is the current through the capacitor of Equation (6.28).

If IEEE 18 is not met, the process may require more than one iteration resizing the size of the capacitor bank.

6.6.4 Test Out Resonant Conditions

Once the filter parameters have been selected, it is important to verify for nonresonant conditions between the capacitor bank of the filter and the inductive reactance of the system. To carry out this task rigorously, a harmonic analysis program is needed to determine the frequency response of the system and to assess whether the desired reduction in harmonic distortion levels is achieved. We show next how far we can go by manually applying the described procedure.

6.7 Example 1: Adaptation of a Power Factor Capacitor Bank into a Fifth Harmonic Filter

Suppose that a capacitor bank installed for reactive power compensation at a six-pulse power converter application is to be tuned to the fifth harmonic. We need to determine the required reactor size and verify whether capacitor bank operation parameters fall within IEEE 18 recommended limits.

Assumed data:

Harmonic current to filter: Fifth.

System phase-to-phase voltage: 13.8 kV.

Power factor capacitor bank size: 4.5 MVAR @ 15 kV.

Plant load: 8 MVA composed of six-pulse static power converters.

Using the preceding methodology, we carry out the following calculations:

Capacitor bank reactance. Typically, the X/R relation for this type of bank is of the order of 5000; therefore, the resistance can be ignored.

$$X_{C1} = \frac{kV_{L-L\,rated}^2}{MVAR_{rated}} = \frac{15^2}{4.5} = 50 \text{ [Ohms]}$$

Calculate the series reactor required. Air core reactors typically have an X/R ratio of the order of 30 to 80. Again, resistance can be disregarded.

From Equations (6.20) to (6.22) we obtain the following:

$$X_{L1} = \frac{X_C}{h^2} = \frac{50}{5^2} = 2.0 \text{ [Ohms]}$$

Determine whether capacitor operating parameters fall within IEEE 18 recommended limits. rms current through the filter:

$$I_1 = \frac{V_{L-N}}{(X_C - X_L)} = \frac{(1.05)\left(\frac{13,800}{\sqrt{3}}\right)}{(50-2)} = 174.3 \text{ [A]}$$

If we assume the harmonic current from the load is inversely proportional to the fundamental current:

$$I_5 = \frac{1}{h}\frac{kVA_{load}}{\sqrt{3}(13.8)} = \frac{1}{5}\left(\frac{8000}{\sqrt{3}(13.8)}\right) = 66.9 \text{ [A]}$$

$$X_{C5} = \frac{X_{C1}}{h} = \frac{50}{5} = 10 \text{ [}\Omega\text{]}$$

Peak and rms voltage through the capacitor:

$$V_{Cpeak} = \sqrt{2}(V_{C1} + V_{Ch}) = \sqrt{2}(X_{C1}I_1 + X_{C5}I_5)$$

$$= \sqrt{2}(50 \times 174.3 + 10X66.9)$$

$$= \sqrt{2}(8715 + 669) = 13,271 \ [V]$$

$$V_{Crms} = \sqrt{(V_{C1}^2 + V_{Ch}^2)} = \sqrt{(8715^2 + 669^2)} = 8741 \ [V]$$

If 8660 V (line-to-neutral voltage for a 15 kV system) capacitors are used, then the capacitor voltage is as follows:

$$\frac{V_{Crms}}{V_{Crated}} = \frac{8,741}{8,660} = 1.009 \ \text{p.u.}$$

(below the 1.1 p.u. limit of IEEE 18).

$$\frac{V_{Cpeak}}{V_{Cpeak \ rated}} = \frac{13,271}{\sqrt{2}(8,660)} = 1.084 \ \text{p.u.}$$

(below the 1.2 p.u. limit of IEEE 18).

The rms current through the reactor is the summation of all rms currents that will flow through the filter. The assumption here is that only the fifth harmonic is involved:

$$I_{Crms} = \sqrt{I_1^2 + I_5^2} = \sqrt{174.3^2 + 66.9^2} = 186.7 \ [A]$$

$$\frac{I_{Crms}}{I_{Crated}} = \frac{186.7}{\left(\dfrac{4500}{\sqrt{3}(15)}\right)} = \frac{186.7}{173} = 1.08$$

(below the 1.35 p.u. limit of IEEE 18).

However, caution should be exercised because a harmonic filter often serves as a sink for currents from adjacent frequencies. Without a harmonic load flow program, it is impossible to be precise about the amount of harmonic currents (other than those for which the filter is tuned) that will flow through it. Thus, in approximate calculations, a factor of 1.15 to 1.2 is sometimes used. In our example, we are within limits, even considering the largest range:

$$Recalculated \ \frac{I_{Crms}}{I_{Crated}} = \frac{186.7}{\left(\dfrac{4500}{\sqrt{3}(15)}\right)}(1.2) = \frac{186.7}{173} = 1.296 < 1.35$$

Reactive power delivered by the capacitor bank is

$$\text{kVAR per phase} = \frac{(V_{C\text{rms}})(I_{C\text{rms}})}{1000} = \frac{\sqrt{(V_{C1}^2 + V_{C5}^2)} \cdot \sqrt{(I_{C1}^2 + I_{C5})^2}}{1000}$$

$$= \frac{(8741)(186.7)}{1000} = 1632$$

Three-phase capacitor power is

$$\text{kVAR}_{3\text{-phase}} = 1632 \times 3 = 4896 = 4.9 \text{ MVAR}$$

The total capacitor output will be derated because the capacitor bank is of a higher-voltage class:

$$\text{MVAR}_{\text{derated}} = \text{MVA}_{\text{rated}} \left(\frac{kV_{L-L}}{kV_{\text{rated}}}\right)^2 = 4.5 \left(\frac{13.8}{15}\right)^2 = 3.8$$

Therefore, considering the influence of the load harmonics, the ratio between delivered and (de)rated power is

$$\frac{\text{kVAR}_{3\text{-phase}}}{\text{kVAR}_{\text{rated}}} = \frac{4.9}{3.8} = 1.29$$

(below the 1.35 p.u. limit of IEEE 18).

Up to this point, we would only need to verify the parallel resonant points. However, because the assumed load does not contain a characteristic harmonic of frequency lower than the fifth harmonic, it really does not matter where the parallel resonant lies.

6.8 Example 2: Digital Simulation of Single-Tuned Harmonic Filters

This example illustrates that harmonic analysis is greatly simplified, on one hand, and that more valuable information is obtained for the assessment, on the other hand, using specialized harmonic analysis software. One of the pieces of information that is extremely helpful in the analysis of mitigating measures for harmonic control is the frequency-dependent plots. These include spectral content (harmonic spectrum) of voltage and current signals and impedance–frequency characteristics of the distribution system before

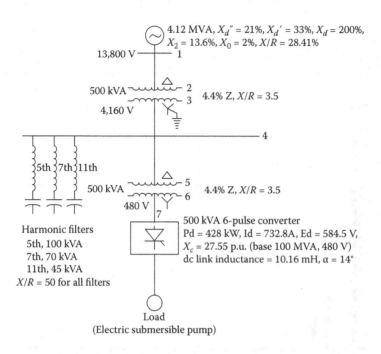

FIGURE 6.8
Electric diagram of an oil offshore installation with a harmonic-producing load.

and after the application of filters. Therefore, it is possible to assess the performance of the system step-by-step with the application of different mitigating methods like, for instance, increasing the number of harmonic filters until desired THD values are reached. This type of analysis is often combined with economical assessment to obtain a comprehensive evaluation harmonic filtering or any other harmonic control scheme.

Figure 6.8 shows a diagram of a typical installation of an electric submersible pump in an oil field offshore platform involving a variable frequency drive fed off from an individual synchronous generator. In these types of installations, in which all or most of the load is nonlinear, it is common to experience large-waveform harmonic distortion due to the lack of linear load components, which act as natural attenuators of waveform distortion. Harmonic filters at the primary transformer that feeds the VFD are tested to show the reduction of harmonic waveform distortion at the generator terminals.

The filter capacitor banks are sized following an inverse relation to their harmonic order. For example, the fifth and seventh harmonic filters are chosen as one-fifth and one-seventh, respectively, of the converter load.

The results obtained for this example are presented in the form of current and impedance vs. frequency diagrams in Figures 6.9 to 6.12.

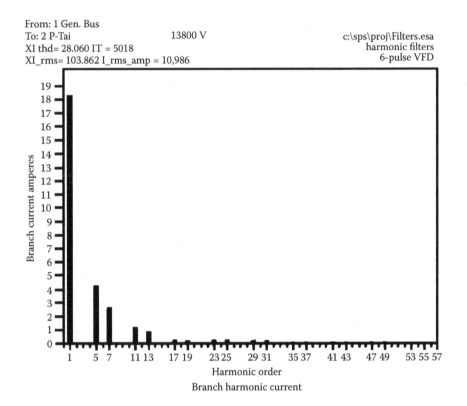

FIGURE 6.9
Current in branch 1–2, with no filter, THD_I = 28.1%.

Figure 6.9 shows the harmonic spectra of the current at branches 1 to 2 with no filters installed. Figure 6.10(a) to (c) describes the effect of a 100 kVA fifth harmonic filter, including branches 1 to 2 and filter currents, as well as the $Z–f$ plot that portrays the series ($Z = 0$) and parallel (just prior to the series) resonant points of the filter. Figure 6.11(a) to (c) shows the same results for a scenario in which fifth and seventh filter branches are applied. Finally, Figure 6.12(a) to (d) shows the effect of additionally including the 11th harmonic branch.

Table 6.3 summarizes the THD under different filtering scenarios. Notice that the THD_V level without any filter at the VFD is already close to the IEEE 519 limit of 5%, while the THD_I is well above the 5% threshold for all power generation equipment. With the installation of the fifth harmonic filter, the

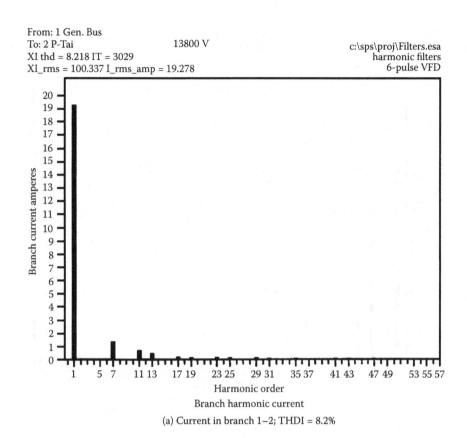

(a) Current in branch 1–2; THDI = 8.2%

FIGURE 6.10
Fifth harmonic filter applied. (*Continued*)

THD_I is reduced to less than half its value when there are no filters, but it is still above the 5% IEEE threshold. If fifth and seventh harmonic filter branches are added, we reach the point at which THD_V and THD_I fall below the recommended limits of IEEE 519.

However, regarding IEEE 18 compliance,[2] the fifth harmonic filter is observed to result slightly above the recommended limits, as observed in Table 6.4. Notice in Figures 6.10(b), 6.11(a), and 6.12(a) how the current through the fifth harmonic filter, which initially comprises a fraction of other harmonics, becomes pure 5th harmonic when 7th and 11th harmonic filters are added. However, when examined in terms of rms values, current remains practically unchanged. A similar behavior is observed for the resultant filter

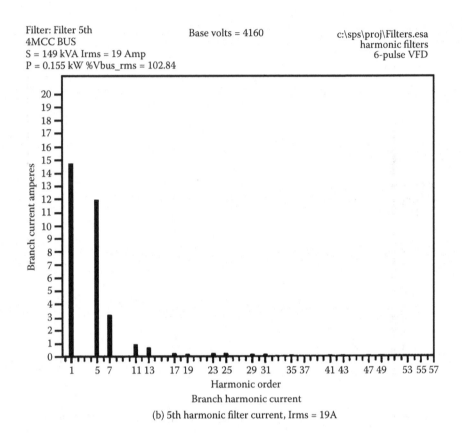

Filter: Filter 5th
4MCC BUS
S = 149 kVA Irms = 19 Amp
P = 0.155 kW %Vbus_rms = 102.84

Base volts = 4160

c:\sps\proj\Filters.esa
harmonic filters
6-pulse VFD

Branch harmonic current

(b) 5th harmonic filter current, Irms = 19A

FIGURE 6.10 (Continued)
Fifth harmonic filter applied. (*Continued*)

kilovolt ampere figures. The current and power values in Table 6.4 are those obtained considering the three filter branches connected. All capacitor peak voltage ratios resulted below the recommended 1.2 limit.

Therefore, a somewhat increased size for the fifth harmonic filter would bring all operating parameters under IEEE 18 compliance.

6.9 Example 3: High-Pass Filter at Generator Terminals Used to Control a Resonant Condition

This example is aimed at illustrating the reduction of THD_V using a high-pass filter in an installation similar to that of the oil company offshore platform of Example 2. The example describes another real-world application

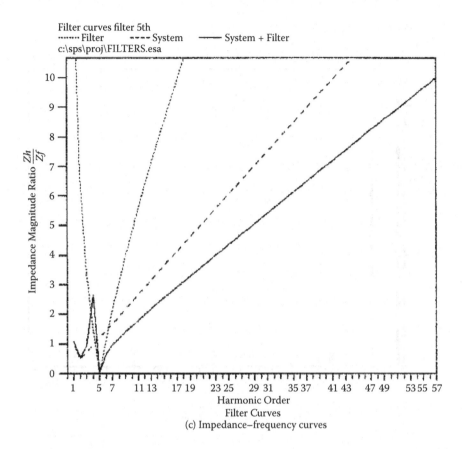

Filter curves filter 5th
········ Filter   ━━━ System   ━━━━ System + Filter
c:\sps\proj\FILTERS.esa

Harmonic Order
Filter Curves
(c) Impedance–frequency curves

FIGURE 6.10 (Continued)
Fifth harmonic filter applied. (*Continued*)

in which harmonic-related problems arose and even damaged a generator unit. This case presented an excellent opportunity for conducting investigations on a complicated phenomenon. The AC source, a 3 MVA synchronous turbo generator, was the power supply for two 1.5 MVA step-down transformers, which in turn fed a number of VFDs powering down-hole electrosubmersible pumps. Apart from a small service transformer that sourced the platform services, the VFDs were the only loads, as illustrated in Figure 6.13.

A number of harmonic measurements indicated that voltage and current harmonic levels were excessively high. The suspecting element was a 0.27 μF surge protection capacitor bank, which apparently combined with connecting cables between generator and transformers to excite a parallel resonant condition at the generator bus. Through additional measurements and

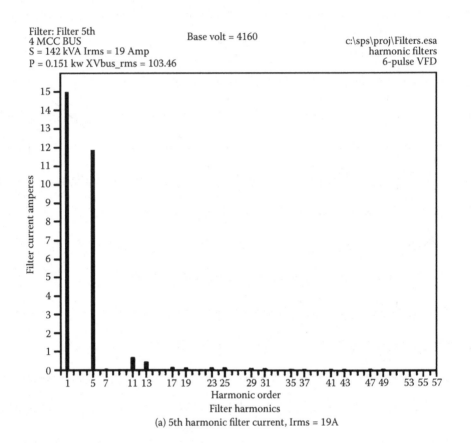

(a) 5th harmonic filter current, Irms = 19A

FIGURE 6.11
Fifth and seventh harmonic filters applied. (*Continued*)

simulation work, this was further confirmed to be the case. Here, we will show the resonant condition and how it was controlled using a high-pass filter to reduce harmonic distortion levels at the generator terminals within IEEE 519 limits.

Figure 6.14 shows the abrupt increase in impedance (dotted line) revealing a parallel resonant condition around harmonics 39 through 43. Such a situation imposed severe stresses on the generator that comprised intense shaft vibration and increased operation temperature. Also shown is the Z–f characteristic of the band-pass filter that was applied at the 600 V generator bus. Filter elements were selected to attain a corner

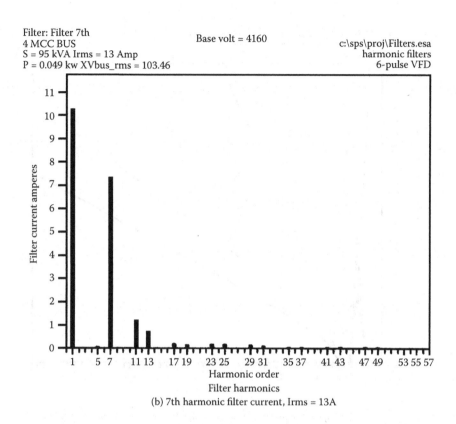

Filter: Filter 7th
4 MCC BUS
S = 95 kVA Irms = 13 Amp
P = 0.049 kw XVbus_rms = 103.46

Base volt = 4160

c:\sps\proj\Filters.esa
harmonic filters
6-pulse VFD

Harmonic order
Filter harmonics
(b) 7th harmonic filter current, Irms = 13A

FIGURE 6.11 (Continued)
Fifth and seventh harmonic filters applied. (*Continued*)

frequency centered at around the 11th harmonic. Notice the system response with the high-pass filter showing a substantial reduction of the impedance around the resonant region.

Figure 6.15 shows how the impedance–frequency characteristic looks at the primary of one of the downstream transformers feeding a VFD at one of the oil wells. At these locations, 5th, 7th, and 11th harmonic filters were installed; this is noticeable on the Z–f characteristic, which also reflects the band-pass filter effect on the distribution system as seen from that location.

Thus, the installation of a low-cost high-pass filter (involving a 67 kVA capacitor bank) at the generator bus allowed a very annoying condition to be brought under control. Interestingly, the parallel resonant phenomenon,

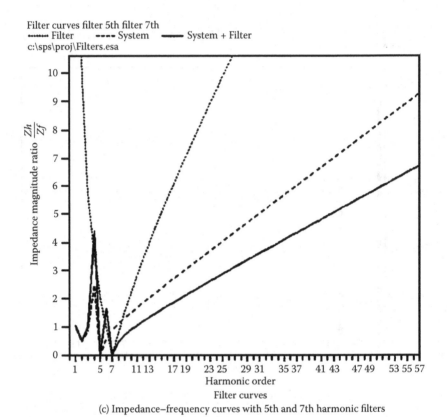

(c) Impedance–frequency curves with 5th and 7th harmonic filters

FIGURE 6.11 (Continued)
Fifth and seventh harmonic filters applied.

as observed in Figure 6.14, involved a frequency range where characteristic harmonics show very small (but apparently strong enough) values that, after undergoing amplification, made THD levels soar.

It is important to mention that applications of high-pass filters at the generator bus are usually combined with the application of single-tuned filters at the VFD locations to get THD levels within recommended limits at the generator terminals, as well as at the individual VFD sites. Leaving the high-pass filter at the generator bus as the only harmonic mitigating method might have caused excessive heating on the high-pass filter elements. In this example, the real intention of the high-pass filter was to achieve a substantial reduction of the large impedance around the parallel resonant region more than providing a low-impedance path to all harmonic currents created at the various VFD sites and making their way to the generation bus.

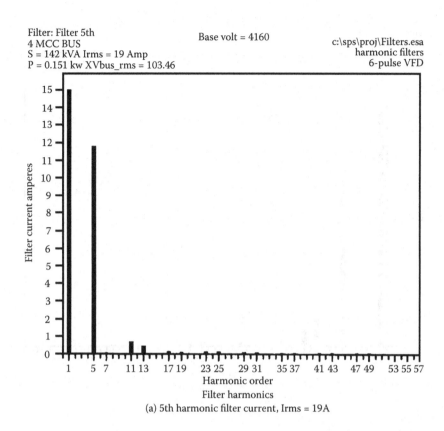

Filter: Filter 5th
4 MCC BUS
S = 142 kVA Irms = 19 Amp
P = 0.151 kw XVbus_rms = 103.46

Base volt = 4160

c:\sps\proj\Filters.esa
harmonic filters
6-pulse VFD

(a) 5th harmonic filter current, Irms = 19A

FIGURE 6.12
Fifth, seventh, and eleventh harmonic filters applied. (*Continued*)

6.10 Example 4: Comparison between Several Harmonic Mitigating Schemes Using the University of Texas at Austin HASIP Program[4]

This last example describes the results obtained combining a number of filtering schemes using the University of Texas HASIP program. The program carries out harmonic analysis, assuming a fundamental voltage of 1 p.u. and short distances between loads and generators. Zero-sequence harmonics are excluded in the analysis.

The screens shown in Figures 6.16 through 6.24 are self-contained in describing size and characteristics of generation, linear load, nonlinear

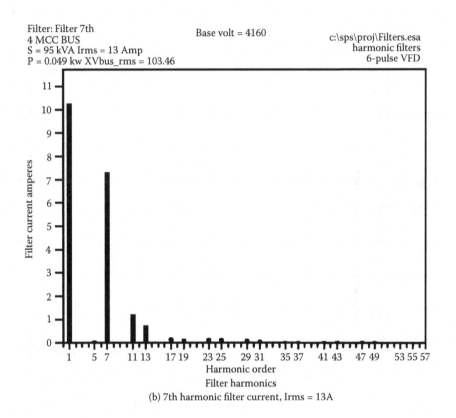

(b) 7th harmonic filter current, Irms = 13A

FIGURE 6.12 (Continued)
Fifth, seventh, and eleventh harmonic filters applied. (*Continued*)

load, capacitor bank, and harmonic filters considered in the analysis. The parameters that were kept fixed are the following:

Generation:

100 MVA base; $X_d'' = 0.20$, $I_{sc} = 500$ MVA

Linear load:

MW: 39

MVAR: 29.3

P.F.: 0.80

Nonlinear load:

Six-pulse converter, 25 MVAR, 09 DPF

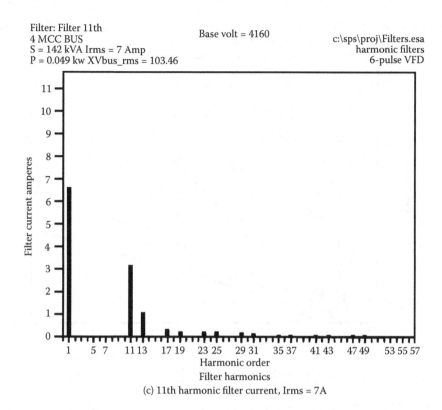

Filter: Filter 11th
4 MCC BUS
S = 142 kVA Irms = 7 Amp
P = 0.049 kw XVbus_rms = 103.46

Base volt = 4160

c:\sps\proj\Filters.esa
harmonic filters
6-pulse VFD

(c) 11th harmonic filter current, Irms = 7A

FIGURE 6.12 (Continued)
Fifth, seventh, and eleventh harmonic filters applied. (*Continued*)

Harmonic filters:

Fifth harmonic: 5 MVAR, $X/R = 50$

Seventh harmonic: 3.6 MVAR, $X/R = 50$

Eleventh harmonic: 2.2 MVAR, $X/R = 50$

Table 6.5 summarizes the different harmonic filtering scenarios tested.

The results indicate the effect of applying different harmonic filtering combinations starting with the application of passive filters of orders 5, 7, and 11. Thereafter, application of individual and different pair sets is tested, and finally, a high-pass filter that reduces harmonic spectral components higher than the 13th harmonic is tested. Figures 6.16 to 6.24 include the waveforms for the capacitor, source current, and nonlinear load current. Also included are the harmonic spectrum of the source and the impedance–frequency characteristics of the system, including filters and capacitor banks. Table 6.5 includes THD values in boldface when they fall in excess of IEEE 519 limits.

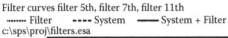

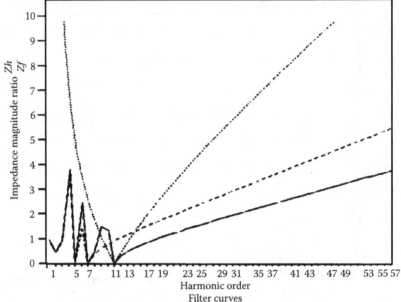

(d) Impedance–frequency curves with 5th, 7th, and 11th harmonic filters

FIGURE 6.12 (Continued)
Fifth, seventh, and eleventh harmonic filters applied.

TABLE 6.3

Voltage and Current Harmonic Distortion at Generator Terminals

Order of Harmonic Filter Tested	No. Filters	5th	5th and 7th	5th, 7th, and 11th
THD$_V$ (%)	4.54	1.9	0.98	0.46
THD$_I$ (%)	28.1	8.2	3.0	1.0

TABLE 6.4

Capacitor Bank Parameters Relative to IEEE 18

Harmonic Filter	Rated Current	rms Current	I_{rms}/I_{rated}	Rated kVA	RMS kVA	kVA$_{rms}$/kVA$_{rated}$
5	13.9	19	1.37[a]	100	142	1.42[a]
7	9.7	13	1.34	70	95	1.21
11	6.2	7	1.13	45	55	0.91

[a] Above IEEE 18 recommended limits.

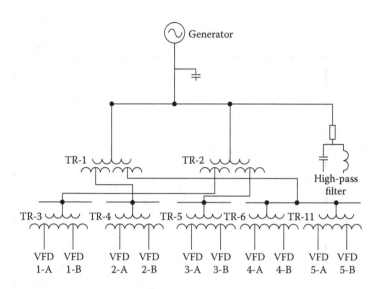

FIGURE 6.13
A high-pass filter at the generator bus on an offshore oil field installation.

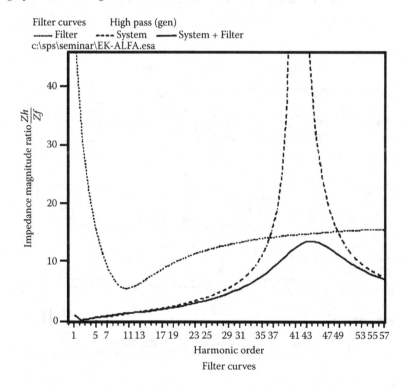

FIGURE 6.14
Impedance vs. frequency characteristics at the generator bus.

c:\sps\seminar\EK-ALFA.esa
Proyecto BEC para el campo EK-BALAN
EK-ALFA

Self impedance: Bus 439 P-TR11

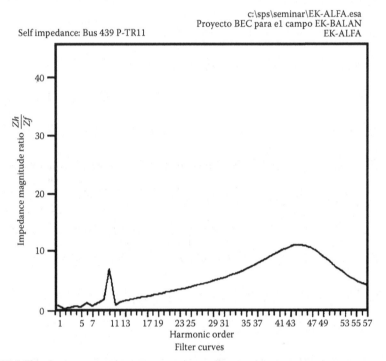

FIGURE 6.15
Impedance vs. frequency characteristics at a downstream VFD.

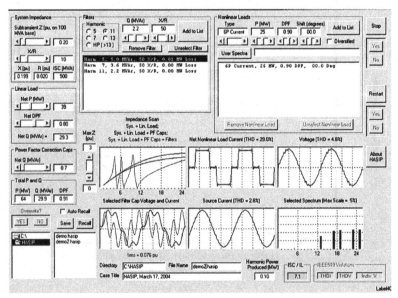

FIGURE 6.16
Effect of 5th, 7th, and 11th harmonic filters, using HASIP. (From HASIP, Version 1, Power Systems Research Group, University of Texas at Austin, Austin, Texas, 2004.)

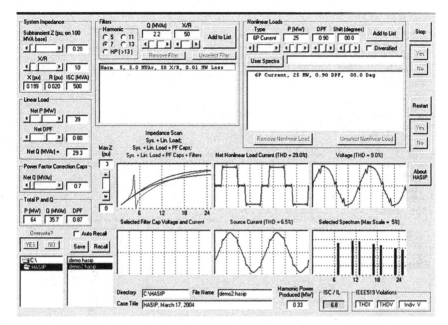

FIGURE 6.17
Effect of 5th harmonic filter, using HASIP. (From HASIP, Version 1, Power Systems Research Group, University of Texas at Austin, Austin, Texas, 2004.)

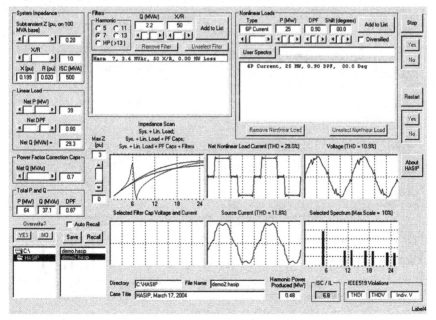

FIGURE 6.18
Effect of 7th harmonic filter, using HASIP. (From HASIP, Version 1, Power Systems Research Group, University of Texas at Austin, Austin, Texas, 2004.)

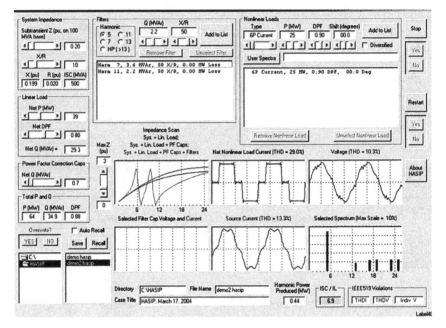

FIGURE 6.19
Effect of 7th and 11th harmonic filters, using HASIP. (From HASIP, Version 1, Power Systems Research Group, University of Texas at Austin, Austin, Texas, 2004.)

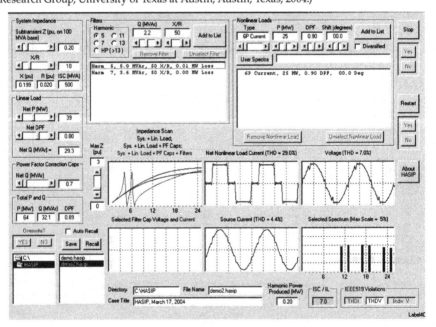

FIGURE 6.20
Effect of 5th and 7th harmonic filters, using HASIP. (From HASIP, Version 1, Power Systems Research Group, University of Texas at Austin, Austin, Texas, 2004.)

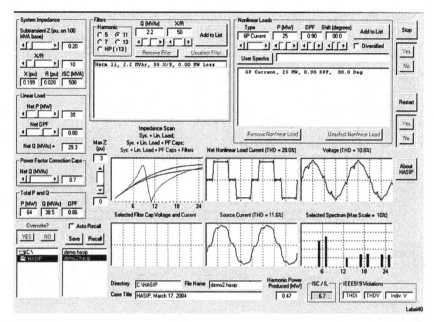

FIGURE 6.21
Effect of 11th harmonic filter, using HASIP. (From HASIP, Version 1, Power Systems Research Group, University of Texas at Austin, Austin, Texas, 2004.)

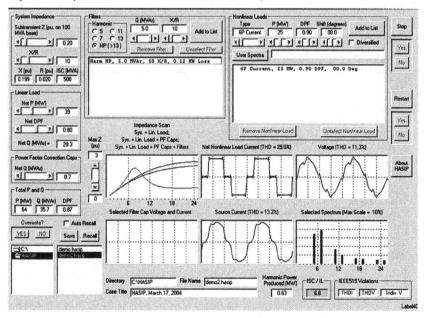

FIGURE 6.22
Effect of high-pass filter to provide a low-impedance path to harmonics above the 13th, using HASIP. (From HASIP, Version 1, Power Systems Research Group, University of Texas at Austin, Austin, Texas, 2004.)

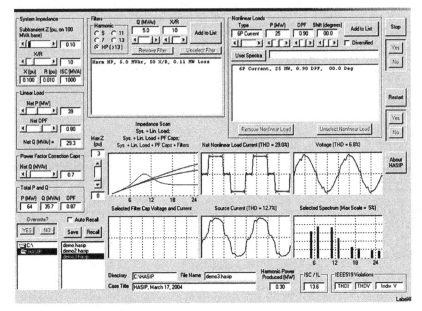

FIGURE 6.23
Effect of filtering harmonics above the 13th combined with a 50% reduction in generator subtransient impedance, using HASIP. (From HASIP, Version 1, Power Systems Research Group, University of Texas at Austin, Austin, Texas, 2004.)

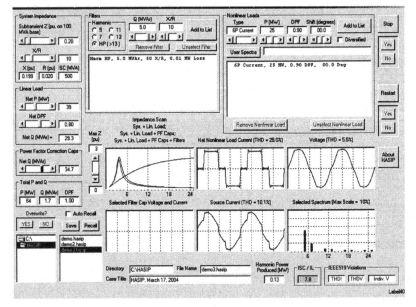

FIGURE 6.24
Effect of filtering harmonics above the 13th combined with an increase of distortion power factor (DPF) to 1.0, using HASIP. (From HASIP, Version 1, Power Systems Research Group, University of Texas at Austin, Austin, Texas, 2004.)

TABLE 6.5

Synopsis of Passive Filter Analysis Using HASIP

NET DPF NET	Q MVAR	I_{sc}/I_L	Harmonic Filter of Order 5	7	11	>13	THD$_I$ Source	THD$_V$ Capacitor	THD$_I$ Converter	Figure No.
0.91	0.7	7.1	X	X	X	—	2.80%	4.80%	29%	6.16
0.87	0.7	6.8	X	—	—	—	6.50%	9.00%	29%	6.17
0.87	0.7	6.8	—	X	—	—	11.80%	10.90%	29%	6.18
0.88	0.7	6.9	—	X	X	—	13.30%	10.30%	29%	6.19
0.89	0.7	7	X	X	—	—	4.40%	7%	29%	6.20
0.86	0.7	6.7	—	—	X	—	11.60%	10.80%	29%	6.21
0.87	0.7	6.8	—	—	—	X	11.30%	13.20%	29%	6.22
0.87	0.7	13.6	—	—	—	X	12.70%	6.80%	29%	6.23
1	34.7	7.8	—	—	—	X	10.10%	5.50%	29%	6.24

Note: Results are plotted in the figures indicated in the last column.

Source: Harmonics Analysis for Ships and Industrial Power Systems (HASIP), Version 1, Power Systems Research Group, Department of Electrical & Computer Engineering, University of Texas at Austin, Austin, Texas, March 17, 2004.

Although the results show the expected outcome of the exercise, the inclusion of the interface screen figures illustrates how useful it is to obtain all the information displayed in the process to determine what combination of filters can make harmonic distortion levels fall within recommended limits. It also helps in understanding the role that every element in the network plays in reaching the desired objective.

6.11 Active Filters

Active filters are an option to the single-tuned filters. They act to clean the distorted voltage and current waveforms by injecting the required amounts of power of specific amplitudes and frequency back into the system to counteract the spikes and notches produced by single or multiple harmonic currents. They can therefore compensate reactive power, and reduce or eliminate lamp flicker produced by subharmonic currents. In performing all these duties, they serve to lessen voltage unbalances in the power system.

Active filters do not pose a risk to engage in resonance conditions with the electric network and are not impacted by changes in the network configuration.

Figure 6.25 shows a PSCAD model representing a six-pulse STATCOM-based active filter with control system as described in Fujita and Akagi.[5]

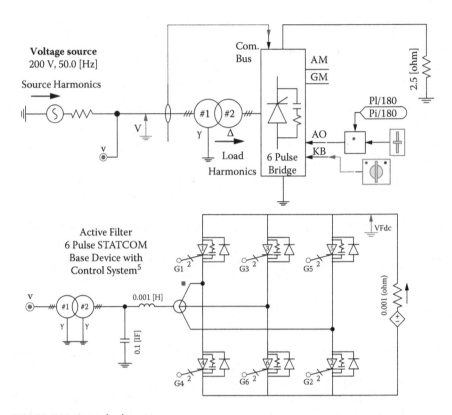

FIGURE 6.25 (See color insert.)
PSCAD model of a STATCOM-based active filter. (Adapted from Fujita, H., and Akagi, H., A Practical Approach to Harmonic Compensation in Power Systems—Series Connection of Passive and Active Filters, *IEEE Trans. Ind. Appl.*, 17(6), 1020–1025, 1991.)

Figure 6.26 illustrates the source and load harmonics currents, and Figure 6.27 depicts the source and load harmonic spectra, respectively. In Figure 6.26 the filter current is also displayed.

Notice the effectiveness of the active filter acting on all of the load current spectral components drastically reducing harmonic current amplitude in the source current, as illustrated in Figures 6.26 and 6.27.

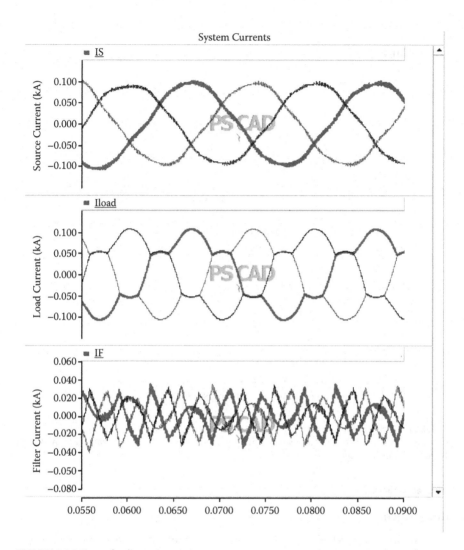

FIGURE 6.26 (See color insert.)
Source, load, and filter currents.

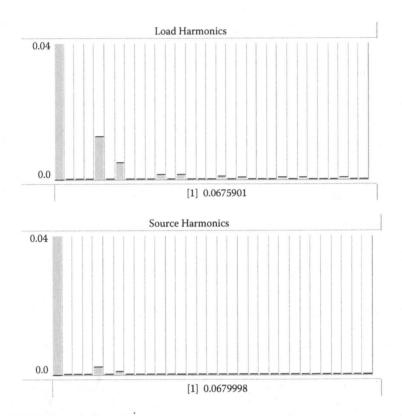

FIGURE 6.27 (See color insert.)
Harmonic spectra of the load and source currents, respectively, from the example shown with the active filter installed.

References

1. ANSI/IEEE 519-1992, *IEEE Recommended Practices and Requirements for Harmonic Control in Electrical Power Systems.*
2. IEEE 18-2002, *IEEE Standard for Shunt Power Capacitors.*
3. The Mathworks, Three-Phase Harmonic Filter, http://www.mathworks.com/access/helpdesk/help/toolbox/physmod/powersys/threephaseharmonic-filter.html.
4. Harmonics Analysis for Ships and Industrial Power Systems (HASIP), Version 1, March 17, 2004, Power Systems Research Group, Department of Electrical & Computer Engineering, University of Texas at Austin.
5. Fujita, H., and Akagi, H., A Practical Approach to Harmonic Compensation in Power Systems—Series Connection of Passive and Active Filters, *IEEE Trans. Ind. Appl.*, 17(6), 1020–1025, 1991.

7

Other Methods to Decrease
Harmonic Distortion Limits

7.1 Introduction

The first technique to control harmonic-related problems in industry involved substantial use of single-tuned filters to offer a low-impedance path to harmonic currents. Interestingly, it is not difficult to find harmonic-producing loads in the megavolt ampere range in industry operating with no harmonic filters. This is a difficult issue for power utilities to control because the existing standards are often more of a reference guideline for industry than a regulatory pronouncement. Large harmonic producers, typically in the industrial sector, may be the only producers that adopt harmonic filtering methods to reduce the otherwise multiple disturbances that may arise beyond the metering point and start affecting sensitive equipment and processes. Due to the high cost involved, this is not a common practice in commercial and residential facilities.

Unfiltered harmonic currents are left to spread freely upstream and downstream from the point of common coupling (PCC) following natural laws of propagation. They may reach adjacent installations and sometimes may even make their way to the utility substation. It is then common to see utilities and harmonic-producing customers in a continuous search for alternative methods to handle and hopefully beat elevated harmonic distortion levels.

For example, special application transformers connected to variable frequency drives, and thus highly exposed to harmonic current overheating, are usually specified as special K factor transformer designs. These special types of transformer constructions do not need to be derated to operate in harmonic environments. Type K transformers are basically designed with improved windings and low-loss iron cores that reduce the amount of additional heating produced by harmonic currents. Notice that harmonic currents on the source side of the converter are by no way controlled or eliminated in the windings. Some cancellation of harmonics can take place, for instance, in phase shift transformers that provide 30° shifting between two six-pulse converters: one fed from delta-connected and the other from wye-connected secondary windings of the transformer.

This chapter describes some of the techniques used in industry to control the flow of harmonic currents produced by nonlinear loads in power systems. The most relevant are the following:

Network reconfiguration

Increase of the short-circuit current ratio

Static multipulse power converters with phase shift transformers

Series reactors

Phase load balancing

Load grouping

7.2 Network Topology Reconfiguration

One measure often advantageous to reduce the effect of unfiltered harmonics is the reconfiguration of the network. Here it is necessary to identify users and sectors in the installation that introduce large amounts of harmonic currents to the system and to characterize its frequency content. As often occurs in residential installations, redistribution of loads by type of loads or across separate circuits can provide an economic solution to drastically reducing disturbances.

Placing the largest nonlinear loads in one or several separate feeders, just as balancing single-phase loads in three-phase systems, would be beneficial. This measure would reduce the otherwise excessive voltage drops from harmonic currents carried through a single path.

If harmonic filters are not an option to consider, mixing linear and nonlinear loads on a feeder may allow the reduction of harmonic distortion because linear loads act as natural attenuators of parallel resonant peaks, as discussed in Chapter 8. This measure should not be contemplated when linear loads comprise sensitive electronic or industrial processes, which may be disrupted if total harmonic distortion (THD) at some point is somewhat increased.

7.3 Increase of Supply Mode Stiffness

Increased ratios between the available short-circuit current and the rated load current make a stronger supply node. This happens whenever power utilities increase their substation's size. It also occurs when industrial customers add some cogeneration on the supply bus to support operation during peak demand.

Stiff AC sources increase the available short-circuit current, for which the ratio between short circuit and load currents is often used as a measure of source stiffness. Strong supply nodes can better absorb transient disturbances in the network and attenuate the effects of large transformer inrush currents, cable energization, and start of large motor loads. The same applies for harmonic currents reaching the substation. The reason for this is that the lower impedance of a stiff supply produces smaller voltage drops, not only for steady state, but also for higher-frequency currents.

High short-circuit currents are then associated with low-impedance sources, which are in turn inverse functions of transformer size. This can be illustrated by calculating the change of impedance when an old transformer of rating MVA1 is replaced by a new transformer rated MVA2. By using the fundamental expression for transformer impedance described in Equation (7.1),

$$X_{TRANSF} = \frac{kV^2}{MVA} \times \frac{X_{leakage}}{100} \tag{7.1}$$

we arrive at the following expression that describes the reactance ratio between transformers rated MVA2 and MVA1, respectively:

$$\frac{X_{MVA2}}{X_{MVA1}} = \frac{\dfrac{kV_2^2}{MVA_2} \times \dfrac{X_{leakage2}}{100}}{\dfrac{kV_1^2}{MVA_1} \times \dfrac{X_{leakage1}}{100}} \tag{7.2}$$

If we assume all other parameters the same, the impedance ratio in Equation (7.2) reduces to

$$\frac{X_{MVA2}}{X_{MVA1}} = \frac{\dfrac{1}{MVA_2}}{\dfrac{1}{MVA_1}} = \frac{MVA_1}{MVA_2} \tag{7.3}$$

That is, the impedance ratio of a new to an old transformer varies with the inverse ratio between the old and the new transformers' megavolt amperes. For instance, a 30 MVA transformer would present an impedance twice as small as a 15 MVA transformer of the same voltage class and a two times increase in short-circuit current, assuming the two of them have the same leakage impedance. In other words, the rating of a distribution transformer for a given voltage can be used as an indication of source stiffness.

At harmonic frequencies, inductive and capacitive impedances of the system vary as a function of frequency, as was shown in Chapter 6:

$$X_{Lh} = h\omega_L \tag{7.4}$$

$$X_{Ch} = \frac{1}{(h\omega_C)} \tag{7.5}$$

A stiffer source will primarily influence the inductive component of the system. Harmonic currents will produce voltage drops affected by the inductive reactance of the system, which is composed by feeder and substation components. In the case of short feeders, the source impedance will be the dominant component. In these situations, harmonic currents are likely to reach the substation and develop voltage drops across the source impedance. In a stiffer system these voltage drops and therefore harmonic distortion will always be smaller due to its smaller impedance.

7.4 Harmonic Cancellation through Use of Multipulse Converters

One-phase converters are used in small load applications. For lower initial costs, half-wave rectifiers can be applied when current requirements are small. Half-wave rectifying produces a DC component that saturates transformers. To limit the former, the use of full-wave rectifying converters is recommended.

The basic polyphase converter is a six-pulse unit. Theoretically, the 12-pulse unit shown in Figure 7.1(a) will eliminate the lower-order harmonics (5th and

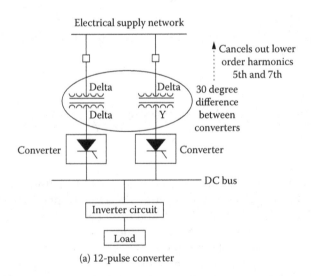

(a) 12-pulse converter

FIGURE 7.1
Phase shift transformer connections for 12- and 24-pulse converters. (*Continued*)

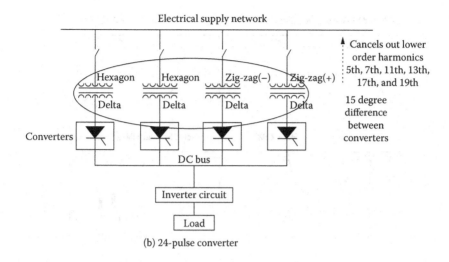

(b) 24-pulse converter

FIGURE 7.1 (Continued)
Phase shift transformer connections for 12- and 24-pulse converters.

7th), for which the first harmonics that will show up are the 11th and 13th. Because the 17th and 19th are not characteristic harmonics, the following harmonic pair to appear will be the 23rd and 25th.

Through additional phase multiplication, it is possible to reduce other harmonic currents. For instance, a 24-pulse unit is built up from four 6-pulse rectifier bridges, each of which has a phase shift of 15° relative to the other rectifying units. This is attained by using phase shifting transformers with separate additional windings connected in zigzag or in polygon, as illustrated in Figure 7.1(b).

If a six-pulse unit were out of service, some cancellation would still be established with two of the six-pulse units 15° out of phase to one another. However, the third unit would show all of the harmonics typical of a six-pulse converter in the system.

The conditions for eliminating harmonics on a six-pulse rectifier composed of N sections using the phase multiplication approach are as follows:

The transformers involved are all of the same transformation ratio and have similar leakage impedances.

The load is split in like parts among the converters.

The firing angle is the same in all converters.

The phase difference between transformers is $60/N$ electrical degrees.

The characteristic harmonics of this harmonic reduction scheme can be expressed as follows:

$$h = kq \pm 1 \tag{7.6}$$

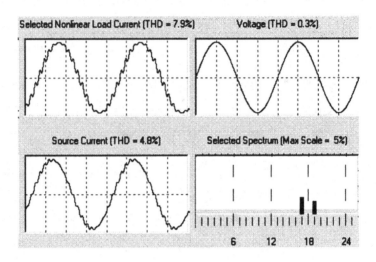

FIGURE 7.2
Eighteen-pulse rectifier.

where h is the harmonic order, q is equal to $6 \times N$, N is the number of six-pulse rectifiers, and k is an integer number (1, 2, 3, …).

If two sections of the rectifier are not equal, noncharacteristic harmonics will always be present as far as the preceding requirements are not met.

Using Equation (7.6), 12-pulse converters would produce cancellation of all harmonics below the 11th, and 18-pulse converters would cancel all harmonics below the 17th.

Figure 7.2 describes voltage and current waveforms at load and source locations, along with the harmonic spectrum of the 18-pulse converter.

7.5 Series Reactors as Harmonic Attenuator Elements

Series reactors have been used in industry for a long time as a way to provide some control on short-circuit current levels. We see them in iron and steel or smelting plants and in power substations or neutral-to-ground connection of generators or power transformers. Series reactors are to some extent also used as harmonic attenuators in industrial applications. Typically, 5% impedance reactors installed on the source side of power converters are seen in a number of applications.

As an energy storage device that opposes the rapid variations of current, a series reactor theoretically provides a two-way attenuation to surge and harmonic currents generated on either side of it. This means attenuation of harmonic currents from the converter (or any other nonlinear load) toward the AC source and of harmonic currents from adjacent customers or from surges generated in the distribution system toward the converter. This looks attractive as a way to provide some relief to transient or subtransient types of events on the power line side created during switching of capacitor banks or long cables or transient disturbances created during line faults, in addition to the attenuation of harmonic currents.

7.6 Phase Balancing

Some electric power companies use four-wire distribution systems with a primary grounded wye and single-phase transformers supplying phase-to-ground voltage to single-phase loads such as residential installations, municipal street lighting, etc. Variations in single-phase loads can create unbalanced currents in three-phase conductors, producing dissimilar voltage drops in the three phases and giving rise to phase-to-phase voltage unbalance. Maximum phase-to-phase or phase-to-ground voltage unbalance may be more critical at the far end of a distribution feeder, where voltage may have experienced a substantial drop during heavy load conditions—particularly in the absence of appropriate voltage profile compensation measures.

A perfectly balanced system is difficult to attain because single-phase loads are constantly changing, producing a continuous unbalance of phase voltages and eventually causing the appearance of even and noncharacteristic harmonics.

7.6.1 Phase Voltage Unbalance

The simplest method to determine voltage unbalance is by calculating the greatest deviation of the phase-to-phase voltage from the average voltage as follows:

$$\text{Voltage\_unbalance}\,(\%) = \frac{\text{maximum deviation from average voltage}}{\text{average voltage}} \times 100 \qquad (7.7)$$

For example, if a 480 V application shows voltages V_{AB}, V_{BC}, and V_{CA} equal to 473, 478, and 486 V, respectively, with an average voltage of $(473 + 478 + 486)/3 = 479$ V, the voltage unbalance is as follows:

$$\text{Voltage\_unbalance}\,(\%) = \frac{7}{479} \times 100 = 1.46\%$$

The amount of voltage unbalance can also be expressed in terms of the negative sequence voltage:

$$\text{Unbalance\_Voltage\_Factor} = \frac{\text{Negative\_Sequence\_Voltage}}{\text{Positive\_Sequence\_Voltage}} \qquad (7.8)$$

7.6.2 Effects of Unbalanced Phase Voltage

When the unbalanced phase voltages are applied to three-phase motors, they give rise to additional negative sequence currents that will circulate in the motor windings, increasing heating losses. The most severe condition occurs under an open-phase situation.

All motors are sensitive to unbalance in the phase voltage. Certain kinds of motors, like those used in hermetically built compressors in air-conditioned units, are more susceptible to this condition. These motors operate with elevated current densities in the windings due to the aggregate effect of the refrigerant cooling.

When a motor is suddenly shut down by the protective system, the first step consists of determining the cause of the disconnection and checking the operation current after it has been put back into operation, to make sure that the motor is not overloaded. The next step consists of measuring the voltage in the three phases to determine the amount of voltage unbalance. Figure 7.3 suggests that a motor is undergoing overheating when voltage unbalance exceeds 2 to 3% for a full-load operation.

Computer operation can be affected by a voltage unbalance of 2 to 2.5%. In general, one-phase loads ought not be connected to three-phase circuits that

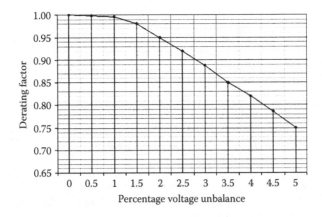

FIGURE 7.3
Derating factor for motors with unbalance in the phase-to-phase voltage. (Adapted from *Optimization of Electric Energy Consumption in Marginal California Oilfields*, EPRI, Palo Alto, California, California Energy Commission, Sacramento, California, 2003.)

provide power supply to sensitive equipment. A separate circuit should be used for that purpose.

Reference

1. *Optimization of Electric Energy Consumption in Marginal California Oilfields*, EPRI, Palo Alto, CA, California Energy Commission, Sacramento, CA, 2003.

8

Harmonic Analyses

8.1 Introduction

Harmonic power system analysis basically requires the same type of information as that required for the analysis of the system under steady-state conditions. The exception to this is the harmonic current source, which must be represented through solid-state switching to re-create the operation of power converters or through appropriate models to represent magnetic core saturation and arc devices. A precise representation of the power system elements will be necessary if an accurate prediction of harmonic response is required.

The propagation of harmonic currents is influenced by a number of factors that relate the offending and affected parties because the two play a major role in the propagation mode of the harmonic currents in the power system. The waveform distortion produced by a strong harmonic source, for example, may still be tolerable to the power system if its dominant harmonic is farther away from natural resonant points in the system. Conversely, a small harmonic source may give rise to large waveform distortion if any of its characteristic harmonics coincides with a resonant frequency in the system, as was illustrated in Example 3 in Chapter 6.

The growing need to conduct harmonic analyses in electrical power systems makes it convenient to review the fundamental principles that govern the flow of harmonic currents. In the process, we must look at relevant aspects like the importance of linear loads as harmonic distortion attenuation elements. It will be important to find the interaction between the different elements of the circuit in relation to the establishment of parallel resonance points. Ascertaining the relationship between the total harmonic distortion level and the voltage notching caused during the operation of thyristors or any other electronic switching devices in power converters is also important.

This chapter describes the most relevant aspects of the study of waveform distortion caused by harmonic currents propagating from their source of origin to the point of common coupling (PCC) and spreading further to adjacent locations and even to the remote AC supply (the utility substation). More than a rigorous mathematical procedure to conduct harmonic analysis, the material described presents an overall depiction of factual features, including the simplified approach to the more elaborate models adopted by modern software programs.

8.2 Power Frequency vs. Harmonic Current Propagation

It is important to highlight that 50/60 Hz power flow studies are centered in the steady-state solution of electric networks to establish optimum operating conditions for a given network to satisfy generation and load requirements. A load flow study will investigate system steady-state load performance under normal operating conditions. Power sources including electric company substations and distributed generation are involved. All significant system loads encompassing resistive and inductive elements, capacitor banks associated with power factor, voltage profile, and harmonic filters are involved in load flow analysis.

Source-equivalent models are often assumed, especially in extended networks. Eventually, industrial installations require thorough representations, particularly when detailed characteristics of distributed generation or network topology are desired. A load flow study is usually carried out in power system analysis to determine voltage, current, and power quantities under steady-state operation.

The reasons for conducting a power flow study are diverse:

Determine the flow of active and reactive power required for estimation of power losses.

Assess the requirements of reactive power compensation.

Estimate voltage profile along the feeders, particularly at remote locations, under heavy load conditions. This helps utilities to define corrective actions to compensate sagging voltage profiles along the feeders and maintain voltage within limits stipulated by voltage regulation policies.

Assess loadability limits of distribution systems under different operation scenarios, which can call for the need of resizing conductors or transformers. Overloaded feeder sectors and transformers contribute to increased losses.

Harmonic flow studies, in contrast, are conducted to determine the propagation of current components of frequency other than the fundamental and the resultant distortion of the voltage waveform. The aim of these studies, among others, is

Determine individual and total harmonic distortion levels produced by nonlinear loads at the location of harmonic sources and at the distribution substation.

Determine harmonic resonant frequencies at capacitor bank locations.

Assess the increased losses due to harmonic currents and take action when they approach thresholds that can have an impact on equipment lifetime.

Specify design characteristic of harmonic filters that can permit the reduction of harmonic distortion levels within recommended limits. This is particularly important when severe harmonic distortion produced by certain customer loads penetrates into adjacent customers' installations.

Properly define size of capacitor banks so that the resultant parallel peak impedance stays away from characteristic harmonics of harmonic-producing nonlinear loads.

Figure 8.1 shows voltage and current waveforms for cases when (a) voltage is in phase with current, (b) voltage leads current, and (c) current leads voltage involving a power factor equal to 0.7. These results are obtained in a typical load flow calculation. Figure 8.2(a) to (c) depicts voltage and current graphs involving third, fifth, and seventh harmonic currents, respectively. The lower graph on each plot is the active power. "Positive watts" refers to power delivered from the source to the load. "Negative watts" refers to power returned from the load storage devices (inductive and capacitive elements) to the source. Notice that the total active power is equal to zero for the three cases in Figure 8.2, which is valid for zero-phase angle between voltage

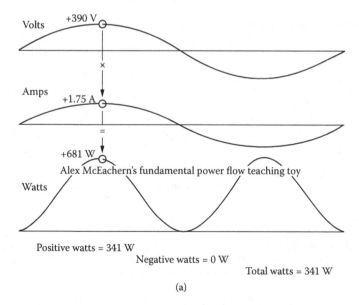

Positive watts = 341 W

Negative watts = 0 W

Total watts = 341 W

(a)

FIGURE 8.1
Voltage and current waveforms in a typical load flow calculation. (a) Voltage in phase with current. (b) Voltage leading current. (c) Current leading voltage. (*Continued*)

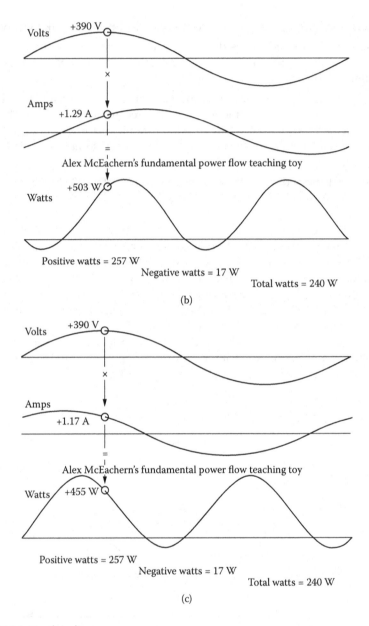

Positive watts = 257 W

Negative watts = 17 W

Total watts = 240 W

(b)

Positive watts = 257 W

Negative watts = 17 W

Total watts = 240 W

(c)

FIGURE 8.1 (Continued)
Voltage and current waveforms in a typical load flow calculation. (a) Voltage in phase with current. (b) Voltage leading current. (c) Current leading voltage.

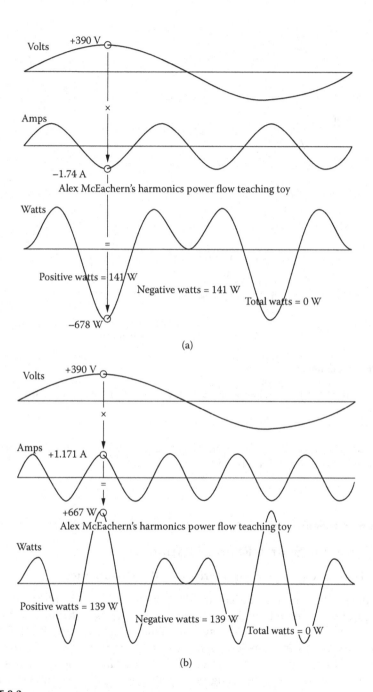

FIGURE 8.2
Voltage and current waveforms containing harmonic currents. (a) Voltage and current with 3rd harmonic. (b) Voltage and current with 5th harmonic. (c) Voltage and current with 7th harmonic. (*Continued*)

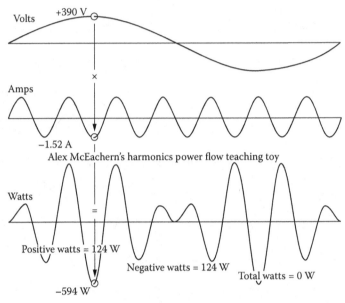

(c) Voltage and current with 7th harmonic

FIGURE 8.2 (Continued)
Voltage and current waveforms containing harmonic currents. (a) Voltage and current with 3rd harmonic. (b) Voltage and current with 5th harmonic. (c) Voltage and current with 7th harmonic.

and current waveforms. These plots were generated with the free harmonic tool from Power Standards Lab.[1]

8.3 Harmonic Source Representation

8.3.1 Time–Frequency Characteristic of the Disturbance

The effects of harmonic sources on the power system will always be more appropriately assessed in the frequency domain, i.e., through a comprehensive Fourier analysis of the system. This requires using manufacturer or measured data to represent harmonic sources from every existing and future nonlinear load in the simulation study. Harmonic current spectra of different harmonic generating equipment or appliances usually include magnitude and phase angle.

The representation of harmonics as ideal current sources assumes that voltages are not distorted. For some nonlinear devices, the representation is considered accurate as long as the real voltage distortion is below around 10%.[2]

Harmonic modeling techniques involve the representation of distortion-producing loads in a form in which they can realistically represent the harmonic sources in the power system network. Harmonic spectra of the load current (see Chapter 2) describe spectral components of individual harmonic sources. In harmonic analysis, these current sources are injected on the electrical system at the point at which they are created, i.e., at the location of the nonlinear load. This is equivalent to superimposing the harmonic currents on the load current waveform.

As described in Chapter 2, there can be multiple sources of harmonic distortion, and every one of them may include different harmonic components. Therefore, harmonic current injection techniques will generally entail the representation of a number of different spectral components that usually fall under one of the following categories:

Power electronic devices

Arc-type devices (electric furnaces, fluorescent lamps, etc.)

Ferromagnetic devices (transformers, induction motors, etc.)

For solid-state devices, characteristic harmonic content can be determined in a straightforward way based on the number of the rectifier steps. For the cases involving arc devices and transformers, it is necessary to obtain the typical harmonic spectra.

Figures 8.3 and 8.4 show typical spectra of 6- and 12-pulse rectifiers in power converters, respectively. They represent the order and amplitude of every harmonic current that will be injected at the location of the nonlinear load.

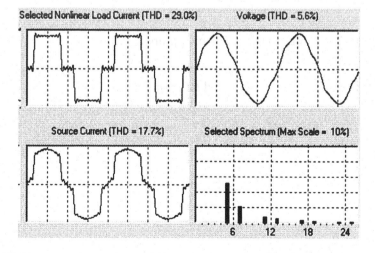

FIGURE 8.3
Six-pulse rectifier.

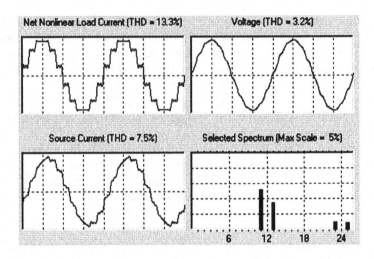

FIGURE 8.4
Twelve-pulse rectifier.

Together with the harmonic spectrum, the nonlinear load current, source current, and power factor correction capacitor voltage waveform are shown. This example was obtained using the unrestricted University of Texas Harmonic Analysis Program (HASIP).[3]

The series decomposition of a signal in its spectral components, as described in Chapter 1, can be carried out using the following formulas:

$$f(t) = \frac{a_0}{2} + \int_{-\infty}^{\infty} (a_n \cos n\omega t + b_n \sin n\omega t) \tag{8.1}$$

where $\omega = 2\pi/T$ and n is the harmonic number, with

$$a_n = \frac{2}{T} \int_{0}^{T} f(t)\cos n\omega t\, dt \tag{8.2}$$

for $n \le 0$ and

$$b_n = \frac{2}{T} \int_{0}^{T} f(t)\sin n\omega t\, dt \tag{8.3}$$

for $n \le 1$.

Harmonic sources are generally dispersed and are usually modeled as current sources of a frequency corresponding to desired harmonic current.

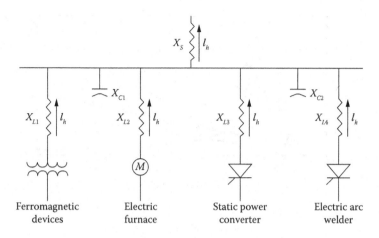

X_{C1}, X_{C2} = Reactance of power factor correction capacitors
$X_{L1} - X_{L4}$ = Step down transformers
X_S = Source impedance

FIGURE 8.5
Harmonic injection current from different sources.

Figure 8.5 portrays a typical model of harmonic current simulation involving a number of sources.

Most software tools include typical harmonic sources so that the user does not need to exercise additional efforts in building them to make an assessment of the problem, particularly during planning stages of a network. In some instances in which harmonic resonance is suspected to be the source of specific disturbances, harmonic current spectra from measurements are preferred. These will provide a far more accurate representation of the harmonic source.

The degree of voltage signal distortion will depend on the amplitude of the harmonic current source and on its propagation on the network. Therefore, an industrial customer may produce harmonic currents that can create some degree of distortion on the voltage waveform at adjacent customer locations. The level of distortion will depend on how much harmonic current will flow toward the source and how much of it will be shared with adjacent facilities.

Inductive and capacitive impedances play an important role in the harmonic current propagation phenomenon. Connecting service drops, transformers, and capacitor banks are some of the elements that can contribute to harmonic current damping or to the excitation of resonant frequencies that can produce significant amplification of voltage distortion.

Impedance scans are used to produce an overall representation of the system response as a function of frequency at specific network locations. This impedance vs. frequency characteristic is generally determined at locations where nonlinear loads, capacitor banks, or harmonic filters exist in

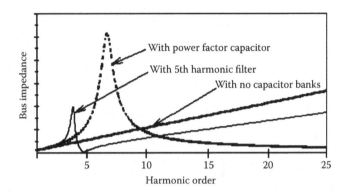

FIGURE 8.6
Impedance vs. frequency response of a typical distribution system.

the network. They can pinpoint capacitor bank resonant conditions like that occurring near the seventh harmonic in Figure 8.6, and they can depict the system response when a fifth harmonic filter is applied, as demonstrated in the same example.

Impedance scans are an excellent tool to anticipate system response in planning network or load expansions.

8.3.2 Resonant Conditions

Natural resonant conditions are important factors that affect the harmonic levels and total harmonic distortion of the system. Harmonic parallel resonance is a large impedance to the flux of harmonic currents; series resonance is a small impedance only limited by a resistive element. When resonant conditions are not a problem, it is because the system has the capacity to absorb important amounts of harmonic currents.

Series harmonic resonance is the result of the series combination of two elements in the electric network seen from the nonlinear load. For instance, a power factor capacitor bank at the primary side of the service transformer feeding a nonlinear load and the inductance of the transformer form a series LC circuit seen from the harmonic current source, as illustrated in Figure 8.7(a). This is actually the operation principle of a single-tuned harmonic filter, in which the series reactor, more than being provided by the transformer reactance, is physically a separate reactor in series with the filter capacitor, as described in Chapter 6.

However, there are instances of series resonance in which unintended harmonic filters are formed at customer installations involving power factor capacitors; these should be further assessed. Equation (8.5) in Section 8.7 can be used to determine the risk for capacitor bank damage if a resonant condition relating to a characteristic harmonic of the load occurs.

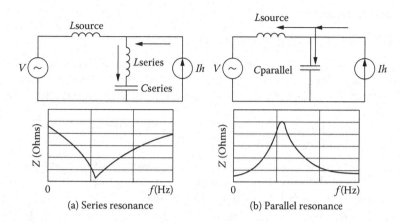

FIGURE 8.7
Series and parallel resonance and its Z–f plot.

Series harmonic resonance presents a low-impedance trajectory to the harmonic currents. The resultant large currents may produce telephone interference in nearby telecommunication systems, conductor heating, and excessive voltage distortion on capacitor banks.

Parallel harmonic resonance (see Figure 8.7(b)) occurs when the reactances of an inductive and a capacitive element, which from the harmonic source appear like parallel elements, become equal at a given frequency. If this frequency coincides with or falls near one of the characteristic harmonics of the load, this harmonic current will oscillate in the form of trapped energy between the inductive and the capacitive elements of the circuit. To the harmonic current, this is equivalent to a large upstream impedance, and as a consequence, a large voltage drop at the frequency of the harmonic current will take place. Overvoltage conditions that can exceed safe operating limits in capacitor bank may develop.

Because inductance and capacitance are intrinsic elements in a power system, series and parallel resonance phenomena are occurrences that, to some extent, are always present in distribution systems.

8.3.3 Burst-Type Harmonic Representation

There are phenomena in power systems operation that produce waveform distortion for short periods. Line-to-ground faults, inrush currents during transformer energization, kick-in of large loads, and motor starting can be mentioned as typical examples of short-term (burst) harmonic distortion. The modeling of these phenomena is similar to the modeling of steady-state harmonic waveform distortion, except that Fourier transform analysis must perform averaging over time windows consistent with the duration of the disturbance.

8.4 Harmonic Propagation Facts

In general, harmonic currents propagating on a power system follow the same physical laws that govern the propagation of low-frequency phenomena. Some relevant factors that play a role in this process are the following:

Location of the injecting harmonic source on the network:

At individual customer facilities

At a large harmonic producer in an industrial/commercial facility where the potential for harmonic currents propagating to adjacent customer installations is identified.

As sources from multiple customers interacting with one another.

Topology of the power system sourcing the harmonic-producing load. As a simple rule, customers fed off from a feeder in the near proximity of the substation are expected to experience harmonic distortion levels somewhat smaller than those of customers fed off downstream from the far end of the feeder. The reason for this is the smaller source impedance, which, in the case of far end customers, is increased by the per-unit length impedance of the feeder.

In connection with the previous point, harmonic source stiffness is important in defining the extent of waveform distortion. Weak systems are associated with a large source impedance, and stiff systems are associated with a small impedance. Therefore, weaker systems will produce larger voltage drops from harmonic currents than stiff systems. This will have an impact on the total harmonic waveform distortion.

Power system capacitance can give rise to series or parallel resonant conditions that can magnify harmonic currents and voltages, for which it should always be considered in harmonic assessment studies.

Distribution system capacitance includes capacitor banks used for power factor correction and for control of voltage profile. Large sections of insulated cables in underground commercial and residential areas can also introduce significant capacitance.

Transmission line capacitance is generally neglected in harmonic propagation studies because the representation of the network rarely extends beyond the distribution substation. However, it may be included in cases involving long transmission lines where more than one large industrial facility with substantial harmonic-producing loads are fed off from the same transmission line.

Linear load effects are important in defining the extent of penetration of harmonic currents into the electric network. Minimum linear loading leads to small attenuation of harmonic distortion and can be used to establish the maximum levels of harmonic distortion that can be expected. Likewise, maximum linear loading means greater attenuation of harmonic distortion and can reveal the degree of minimum level of harmonic distortion that should be expected.

8.5 Flux of Harmonic Currents

How do harmonic currents flow in an electrical network? The easy answer to this question is that harmonic currents freely flow from their source of generation to whatever point electrically connected to them. However, harmonic currents will find it easier to propagate on reduced impedance paths. What this means is that once we apply Fourier decomposition techniques to find the harmonic components of the current, we can treat every single harmonic source as we would a 50/60 Hz current.

The essential difference with power frequency currents, however, is that network impedances are frequency dependent, as described by Equations (7.4) and (7.5) of Chapter 7. The consequence of this is that higher-order harmonics will see a large inductive reactance and a small capacitive reactance. Conversely, lower-order harmonics will propagate through small inductive and large capacitive paths.

This simple fact allows us to understand that higher-order harmonic currents produced at the end of an uncompensated feeder will not have the same penetration toward the distribution substation that lower-order harmonics will. Using the resonant frequency equation,

$$f_{res} = \frac{1}{2\pi\sqrt{LC}} \tag{8.4}$$

we can also realize that a capacitor bank would present a lower resonant frequency when interacting with a large inductive component, e.g., if it were located at the end of a feeder rather than if it were located close to the substation.

We would then find all possible combinations of L and C, but maybe one that is relevant is when L and C are large, which would combine to produce a resonant point on the lower region of the frequency spectrum where the largest characteristic harmonic currents of six-pulse converters, fluorescent lighting, and electric furnaces are found.

Harmonic current propagation thus depends on the electrical parameters of the circuit, in the same way as a fundamental frequency power flow, starting from the point where it is created (the location of nonlinear loads) and spreading upstream throughout utility source impedance elements, which are generally small. However, harmonic currents may also reach adjacent customer locations where power factor correction capacitors offer low-impedance paths at high frequencies. See Figure 8.8.

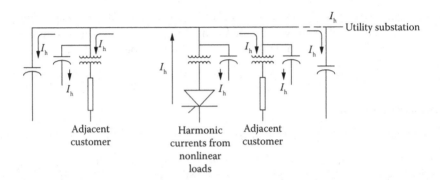

FIGURE 8.8
Harmonic currents and their propagation to distribution networks and adjacent installations.

In the absence of harmonic filters, harmonic currents not only can reach the AC source and produce undesirable harmonic distortion levels, but also may produce transformer overheating and other unfavorable effects, as discussed in Chapter 4.

The penetration of harmonic currents into the electrical network has become a hot issue for electric utilities in the light of stringent regulations calling for improved power quality indices and compliance with industry standards. IEEE 519[3] on harmonic recommended limits is undergoing revision aimed at producing specific electric utility application guidelines.[4,5]

The main concern with harmonic distortion is the steady increase of nonlinear loads at commercial and industrial facilities, including all kinds of power converters in massive personal computer utilization, among others. Figure 8.9 depicts a simplified analysis model for the one-line diagram of Figure 8.8, showing line and transformer impedances. The aim of this illustration is to show how a large harmonic current injected by a particular customer has the potential to penetrate through multiple paths into the distribution network. It can reach adjacent customer installations and can even propagate to the AC power source, where it can create an undesirable increase of harmonic distortion. Notice that the flow of harmonic currents on

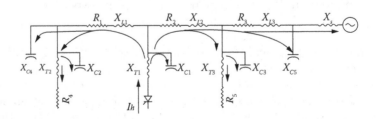

FIGURE 8.9
Harmonic current flow on a typical distribution network.

that part of the feeder between harmonic and AC sources is opposite of the normal flow of power frequency current.

8.5.1 Modeling Philosophy

The AC source is usually represented through a simple Thevenin equivalent using the short-circuit impedance. Some computer modeling programs offer very detailed representation of AC sources, including excitation system in synchronous generators and mechanical and electrical constants. This provides the possibility to incorporate in-depth representations of additional sources of energy in the simulation model. These may include distributed generation used by utilities to cope with voltage and power support during peak demand hours or cogeneration resources in industry.

Transmission lines should be modeled when a number of large industrial customers are fed off from the same transmission line because, due to their lower impedance, harmonic currents are little attenuated and may easily spread to adjacent facilities.

Distribution feeders are represented by their per-unit-length positive sequence impedance, usually comprising resistance and inductance for overhead lines and including capacitance for underground feeders. Harmonic analysis software usually offers a number of power line models, ranging from constant frequency (Bergeron) to frequency-dependent models. In all cases, conductor configuration and soil parameters are needed so that the program can calculate the necessary impedance/admittance matrices.

For industrial loads fed off from dedicated feeders, it is important to set up a detailed representation of the low-voltage components, including linear and nonlinear loads as well as any kind of energy storage devices, like capacitor banks and reactors. On the primary side of the service transformer, a high-voltage side equivalent impedance can be used as a first approach because the transformer is the dominant impedance.

Capacitor banks should be modeled at the primary and secondary sides of the transformer whenever present because they are important elements that can excite resonant conditions. At the primary side, the capacitance of the distribution line is not significant. However, insulated cables of considerable length should be modeled because, as capacitive elements, they may play an important role in establishing the level of telephone interference, generally produced by higher-order harmonics.

8.5.2 Single-Phase vs. Three-Phase Modeling

For most harmonic studies, the three-phase model representation of the system using the positive sequence parameters will be sufficient. Exceptions to this statement may be found in the following cases:

Unbalanced systems. In this case, the unbalanced system or the uneven phase harmonic sources can be incorporated to determine the per-phase harmonic components precisely. If telephone interference is to be assessed, higher-order triplen harmonics (produced, for instance, during transformer saturation) will be of special interest because they are added in phase and can be present at considerable distances from their source.

One-phase capacitor banks. Although rarely found, single-phase compensation on three-phase systems will also require per-phase models.

For one-phase models, the per-phase representation is carried out in separate form for the different propagation modes and thereafter converted to per-phase quantities. This is done through eigenvector analysis.

8.5.3 Line and Cable Models

Most software tools include some form of fixed or frequency-dependent models for transmission lines as described previously. Frequency-dependent quantities are the earth return, which is calculated from Carson's equations,[6] and skin effect, which is generally derived using Bessel functions.[7] For short lines or low-frequency applications, a series impedance may provide a reasonably good representation of the line. However, for underground cables or lines with significant portions of insulated cables or in studies that involve higher-order harmonics, it will be important to include the shunt capacitance in their representation, as depicted in Figure 8.10.

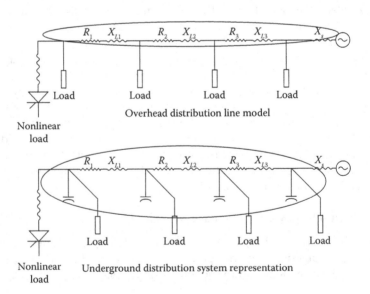

FIGURE 8.10
Model representation of overhead and underground distribution systems.

For multiphase lines, equations are solved using modal transformation as described in reference 8.

For transmission lines, the representation of long lines must consider transpositions and distributed capacitances for a more adequate modeling.

8.5.4 Transformer Model for Harmonic Analysis

Transformers have two components that are of most interest:

Leakage impedance

Magnetizing impedance

Normally, the magnetizing impedance is much larger than the leakage impedance when the transformer in not operated on the saturation region. For harmonic analysis, a representation of the transformer should include a current source[9] and a resistance (Rm) to account for core losses, as depicted in Figure 8.11. For low-order harmonics, a lumped impedance representation for the leakage impedance can be used.

When required by the model used, the X/R ratio can be assumed to have a typical value of 10 and two or three times as high for transmission transformers. If the transformer is not a significant source of harmonics, the magnetizing impedance can be ignored.

8.5.5 Power Factor Correction Capacitors

Capacitor bank placement in distribution or industrial networks will influence harmonic resonance frequency. Another consequence of applying capacitor banks in installations with nonlinear loads is that the total power

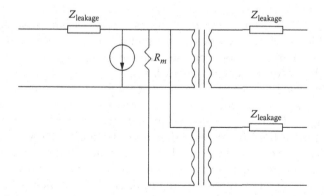

FIGURE 8.11
Transformer model for harmonic analysis. (Adapted from Bonner, A. et al., *IEEE Trans. Power Delivery*, 11(1), 452–465, 1996.)

factor will not be influenced in the same way as the displacement power factor. Real power factor will be smaller than displacement power factor, as described by Equations (1.47) and (1.48) in Chapter 1.

Applying capacitor banks to nonlinear loads can have the effect of increasing waveform distortion on voltage and current signals at the location where capacitor banks are applied as well as at adjacent installations, for which in some specific cases it may be important to extend the analysis to neighboring facilities to get a complete picture of harmonic penetration effects. Single-phase capacitor banks in unbalanced systems may produce an increase in noncharacteristic harmonics.

8.6 Interrelation between AC System and Load Parameters

The following elements of the electric power system are important to consider in the study of harmonic propagation:

Step-down transformers

Resistive components of the load

Rotating machine components

Step-down transformers are important at high frequencies because they form a series reactance with the load. System elements that absorb active power, such as resistive and inductive components, can become significant attenuators of harmonic waveform distortion.

Step-down transformers are usually represented by means of their series leakage reactance. Magnetizing current of the transformer is typically a small percentage (normally below 3%) of its rated load for which the magnetizing branch is often omitted in analysis. However, most software tools offer suitable models to include transformer saturation curves when needed. This could be of interest, for example, when investigating zero-sequence currents in delta/wye-connected transformers.

Linear loads are generally represented by passive elements. At the frequency of the lower-order harmonics, transformer leakage reactance is small compared with the resistive impedance of the load. However, at the frequency of the higher-order harmonics, the step-down transformer reactance can become as large as that of the load, so it can be considered that the transformer represents an important attenuator of high-frequency harmonic currents generated by the nonlinear load, which is favorable to the source.

The resistive portion of the load provides damping that affects system impedance in the proximity of the resonant point. Therefore, the higher the loading of the circuit, the lower the resulting impedance near the resonant frequency. The trajectory through the resistance thus offers a smaller impedance at higher load levels. This is the path taken by harmonics when parallel resonance arises. The typical distribution system response with different load levels is shown in Figure 8.12 at the frequency corresponding to the system's parallel resonance. The shifting position of the bars along the *x*-axis indicates different levels of loading.

Electrical motors can be represented by their short-circuit or blocked rotor impedance at harmonic frequencies. The subtransient impedance is more common to represent induction motors. Often an average of the direct and quadrature axes' impedances is used. This impedance does not provide a significant attenuation of the parallel resonant peak. It can, however, slightly change the resonant frequency. Likewise, load variations can also produce a similar resonant frequency shift.

Balanced systems are modeled using their positive sequence impedances. This includes loads and reactive compensation elements. Under these conditions, harmonic currents will have characteristic phase sequences, as discussed earlier in Section 8.3.

For representation of unbalanced systems comprising uneven distribution of loads on the three phases, single-phase capacitor banks, or asymmetrical configurations of power cables, a per-phase representation using positive, negative, and zero-sequence impedances should be used. Results of the analysis may reveal even harmonics.

Large harmonic distortion should be expected when nonlinear loads account for the greatest percentage of the loads in a given application. This can be a

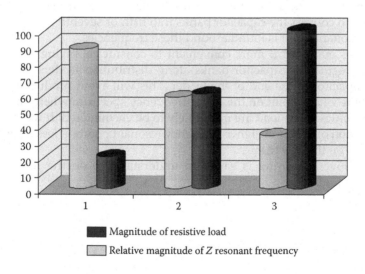

Magnitude of resistive load

Relative magnitude of *Z* resonant frequency

FIGURE 8.12
Typical resistive load effects on the parallel resonant peak impedance.

case of large VFDs fed by dedicated feeders as the only load. Some examples include offshore production platforms in the oil industry and steel mill plants.

Some form of parallel resonance should be suspected when a large harmonic voltage is detected at the source terminals under light load conditions involving a nonlinear load application.

8.6.1 Particulars of Distribution Systems

There might be a variety of types of nonlinear loads. Frequency spectra can usually be obtained from manufacturers. Harmonic analysis software usually includes some generic models for 6-, 12-, 18-, and 24-pulse converters.

Cogeneration is sometimes used in industrial plants such as steel mills and others. Because adding harmonic filters in variable frequency drive installations is likely to convert power factor from lagging to leading, it will be important additionally to check for generator stability. A representation of the machines, including machine constants, is necessary if transient stability or machine response to any other transient condition is assessed.

Nonlinear loads in industry are often cyclic, for which the characterization of harmonic distortion levels may require carrying out long-term measurements to determine the profile of the voltage and current waveform distortion. This is important when harmonic filtering is being considered so that filter components are properly sized to match the maximum load variations.

If we are concerned with harmonic currents making their way to the distribution substation, it will be important to consider that the most common harmonic currents observed at a distribution substation are typically the fifth and the seventh. Existing capacitor banks may be furnished with series reactors to convert the capacitor banks into fifth or seventh harmonic filters if the resonant point is found around those frequencies.

Utilities may find it more convenient to consider applying a number of voltage regulation capacitor banks along distribution feeders instead of a large capacitor at the end of the feeder. This will avoid producing a single parallel resonant point on the lower-frequency spectrum where characteristic harmonics of the load are most likely to be encountered. It would induce the creation of multiple higher-frequency resonant points at less troublesome locations on the spectrum.

The above reasoning is easier to understand if, on the resonant frequency equation,

$$f_{res} = \frac{1}{2\pi\sqrt{LC}}$$

we keep C constant and let L increasingly change.

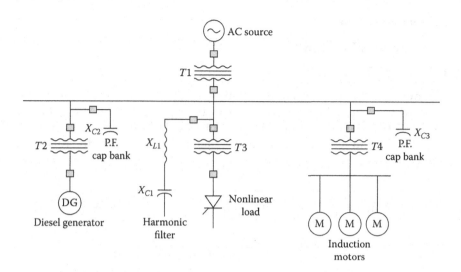

FIGURE 8.13
Typical industrial power system.

8.6.2 Some Specifics of Industrial Installations

As illustrated in Figure 8.13, industrial power systems look like compact distribution systems. Some of them may even have a generation facility running continuously or during peak demand periods to help the power utility cope with voltage regulation problems. However, some important differences must be mentioned:

Capacitor banks often dominate the frequency response of industrial systems and the short-circuit inductance is relatively large.

The parallel resonance is often found at low-order harmonics.

The amount of harmonic-producing loads in industrial systems is often greater than in distribution systems. This is due to the use of large power converters, arc furnaces, variable frequency drives, etc.

Resistive loads are often small, and resonant impedance peaks are not optimally damped near resonant frequencies. This often results in severe harmonic distortion, although to a lesser extent, motor loading does help in providing some attenuation of the resonant impedance peak.

Most industrial installations can be modeled as balanced systems. Loads are generally three-phase and balanced, including harmonic sources, transformers, and capacitor banks.

8.7 Analysis Methods

8.7.1 Simplified Calculations

An oversimplified manual calculation suggested in reference 3 may be used in cases when the system can be represented by the circuit of Figure 8.14. Among the most important calculations for this circuit, we have the system resonant frequency, which is obtained with the following expression:

$$\text{Sys\_res\_freq} = \sqrt{\frac{\text{Short\_circuit\_MVA}}{\text{Cap\_Bank\_MVAR}}} = \sqrt{\frac{\text{Cap\_Bank\_}X_C}{\text{Substation\_Short\_circuit\_}X_L}}$$

$$= \sqrt{\frac{\text{Transformer kVA} \times 100}{\text{Rating of connected capacitors in kVA} \times \text{Transformer } Z\%}}$$

$$(8.5)$$

where Sys_Res_freq is the system resonant frequency expressed as an integer multiple of the fundamental frequency, Short_circuit_MVA is the system short-circuit megavoltamperes at the point under consideration, Cap_Bank_MVAR is the rated megavoltamperes of the capacitor bank at the system voltage, Cap_Bank_X_C is the reactance of the capacitive bank at the fundamental frequency in ohms, and Substation_Short_circuit_X_L is the short-circuit reactance at the supply substation in ohms.

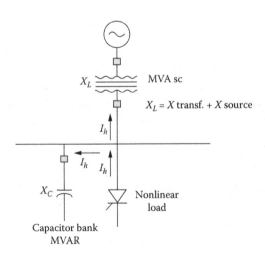

FIGURE 8.14
Circuit for simplified calculations.

If the calculated resonance frequency is close to one of the characteristic harmonics of the source (the nonlinear load), large harmonic overvoltages that may increase waveform distortion to inconvenient levels can develop. A precise and more rigorous study is then required.

In a simplified manner, the system impedance expressed in the frequency domain can be approximated by

$$Z(\omega) = \frac{(R + j\omega L)}{[1 - \omega^2 (LC) + j\omega RC]} \tag{8.6}$$

where $Z(\omega)$ is system impedance as a function of frequency ($\omega = 2\pi f$).

The harmonic voltage at every individual harmonic frequency can then be obtained using:

$$V_h = (I_h)(Z_h) \tag{8.7}$$

where I_h is the current source at the characteristic harmonic h.

Taking into account the contribution of every harmonic considered in the assessment, the rms voltage can be determined from

$$V_{RMS} = \sqrt{V_1^2 + \sum_{h=2}^{n} V_h^2} \tag{8.8}$$

Total harmonic distortion (THD) and the telephone influence factor (TIF) can also be estimated through a simplified approach using Equations (1.39) and (1.42) in Chapter 1. More rigorous calculations can be carried out using individual harmonic voltages and currents.

8.7.2 Simulation with Commercial Software

When circuits become complex, it is necessary to conduct computer simulations, using software capable of carrying out the following calculations:

The frequency response of the system

The frequency response for multiple harmonic sources

The solutions for unbalanced polyphase circuits

The most common method employed in harmonic analysis programs is to make a direct solution of the matrix impedance at multiple frequencies. With this type of solution, nonlinear devices are modeled as ideal current sources at harmonic frequencies. Frequency dependence of the system elements (transmission lines, transformers, motors, etc.) is generally included in calculations. However, the system is considered linear at every individual frequency.

The iterative Newton–Raphson method is commonly used for the solution. The implementation of this solution is done for balanced systems and is generally applicable to the analysis of transmission systems and in distribution networks.

8.8 Examples of Harmonic Analysis

To illustrate the versatility of modern computer simulation software, a couple of cases are presented to illustrate the harmonic analysis on the waveform distortion produced during transformer energization (inrush current) and during a single-phase fault to ground. These cases are chosen to complement other harmonic analysis cases shown throughout the book.

8.8.1 Harmonic Current during Transformer Energization

This example is taken from the PSCAD User's Group home page.[10] It calculates the inrush current on closing the transformer breaker near the zero crossing of the breaker. Figure 8.15 shows the electric diagram of the source and the transformer. Transformer data are provided in Table 8.1.

Figure 8.16 shows plots for the current, voltage, and flux of the transformer when the transformer breaker is closed near the zero crossing of the voltage. Using the Electrotek Concepts TOP,[11] FFT analysis is applied to the transformer inrush current and shown in Figure 8.17. Here we can see that the dominant 60 Hz-based harmonics are the second, third, and fifth, and the calculated total harmonic distortion reaches a level of 103%.

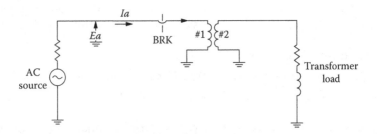

FIGURE 8.15
Circuit diagram of the model used to study transformer inrush current.

TABLE 8.1

Transformer Data Including Saturation Curve Used to Test Inrush Current

Transformer Name	T1	Magnetizing Current at Rated Voltage	
Transformer MVA	1 (MVA)	Enable saturation	1
Primary voltage (rms)	1 (kV)	Point 1—current as a percent of rated current	0.0 (%)
Secondary voltage (rms)	1 (kV)	Point 1—voltage in p.u.	0.0 (p.u)
Base operation frequency	60 (Hz)	Point 2—(I-V)	0.1774 (%)	0.324129 (p.u.)
Leakage reactance	0.010 [p.u.]	Point 3—(I-V)	0.487637 (%)	0.61284 (p.u.)
No load losses	0.0 (p.u.)	Point 4—(I-V)	0.980856 (%)	0.825118 (p.u.)
Copper losses	0.0 (p.u.)	Point 5—(I-V)	2 (%)	1.0 (p.u.)
Model saturation?	Yes	Point 6—(I-V)	3.09543 (%)	1.08024 (p.u.)
Tap changer winding	None	Point 7—(I-V)	6.52348 (%)	1.17334 (%)
Graphics display	Windings	Point 8—(I-V)	20.357 (%)	1.26115 (p.u.)
		Point 9—(I-V)	60.215 (%)	1.36094 (p.u.)
		Point 10—(I-V)	124.388 (%)	1.49469 (p.u.)

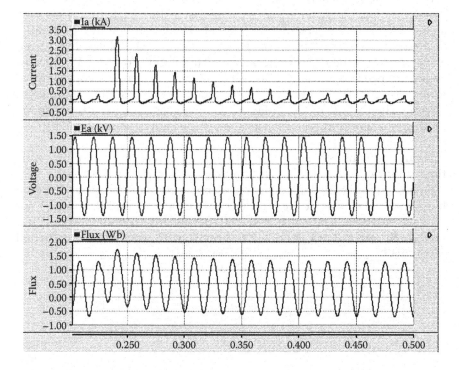

FIGURE 8.16
Current, voltage, and flux waveforms during transformer energization.

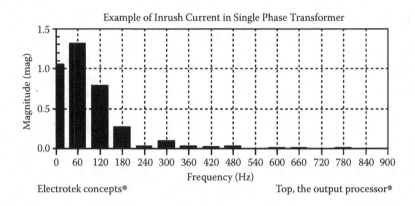

FIGURE 8.17
Inrush current harmonic spectrum.

8.8.2 Phase A to Ground Fault

This example considers a single-phase-to-ground fault at a location F2 (between circuit breaker B1 and transmission line in Figure 8.18). Here again, this is not the type of steady-state harmonics found in the operation of a power converter, but serves well to describe the voltage and current waveforms and their spectral content during a fault to ground. Protective engineers need to consider the distortion imposed on the current waveform during short-circuit conditions for adequate relay setting of protective devices. The PSCAD model in Figure 8.18 includes a long transmission line linking two substations and the location of the fault at point F2 downstream of circuit breaker B1.

Figures 8.19 and 8.20 show the voltage and current on every phase at the time of the fault at the locations of breakers B1 and B2, respectively.

Finally, Figure 8.21 depicts the harmonic spectra of voltage and current harmonic waveforms at the locations of breakers B1 and B2 after B1 and B2 have reclosed. The calculated THD levels are around 10 and 45% at breaker B1 and around 10 and 43% at breaker B2, for voltage and current, respectively.

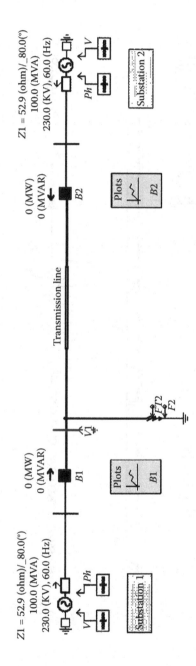

FIGURE 8.18
Model used to calculate harmonic content during a fault to ground.

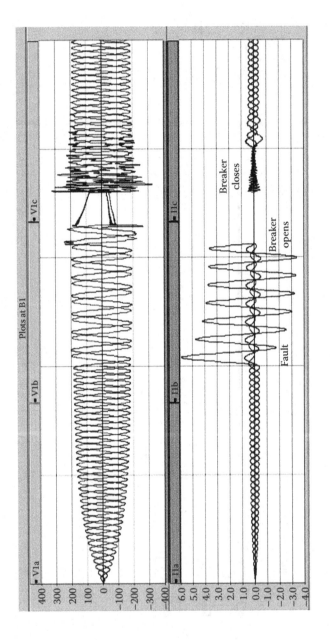

FIGURE 8.19
Voltage and current waveforms during the simulated phase A to ground fault at location B1.

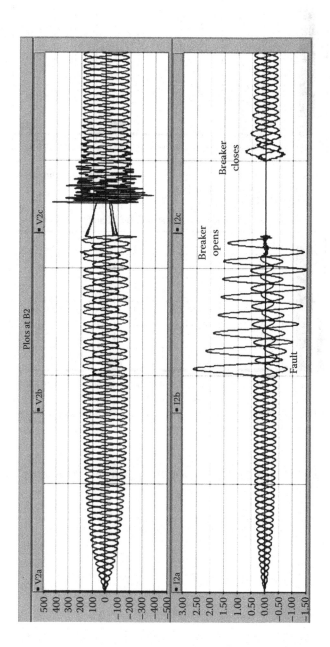

FIGURE 8.20
Voltage and current waveforms during the simulated phase A to ground fault at location B2.

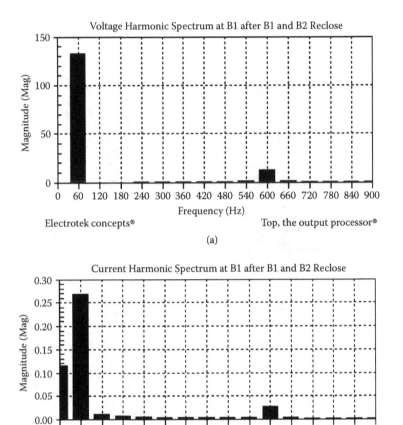

FIGURE 8.21
Voltage and current spectra at locations B1 and B2 after the two breakers have reclosed.
(*Continued*)

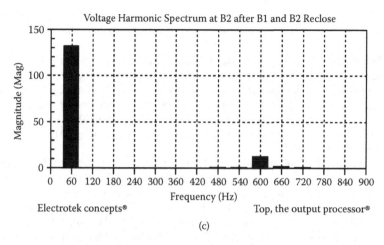

Voltage Harmonic Spectrum at B2 after B1 and B2 Reclose

Electrotek concepts®   Top, the output processor®

(c)

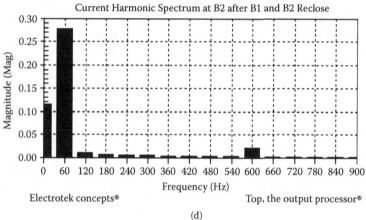

Current Harmonic Spectrum at B2 after B1 and B2 Reclose

Electrotek concepts®   Top, the output processor®

(d)

FIGURE 8.21 (Continued)
Voltage and current spectra at locations B1 and B2 after the two breakers have reclosed.

References

1. McEachern, A., *Power Quality Teaching Toy*, Edition, 2.0.2, Power Standards Lab website, http://www.PowerStandards.com.
2. IEEE 519-1992, *Recommended Practices and Requirements for Harmonic Control in Electric Power Systems*.
3. Harmonics Analysis for Ships and Industrial Power Systems (HASIP), Version 1, March 17, 2004, Power Systems Research Group, Department of Electrical & Computer Engineering, University of Texas at Austin.

4. Halpin, M., Comparison of IEEE and IEC Harmonic Standards, in *Proceedings of 2005 IEEE Power Engineering Society General Meeting*, San Francisco, June 12–16, 2005.
5. Halpin, M., Harmonic Modeling and Simulation Considerations for Interharmonic Limits in the Revised IEEE Standard 519:1992, in *Proceedings of 2005 IEEE Power Engineering Society General Meeting*, San Francisco, June 12–16, 2005.
6. Galloway, R.H., et al., Calculation of Electrical Parameters for Short and Long Polyphase Transmission Lines, *Proc. IEEE*, 111(12), 2051–2059, 1964.
7. Magnunson, P.C., *Transmission Lines and Wave Propagation*, Allyn and Bacon, Boston, 1965.
8. Dommel, H.W., *Electromagnetic Transients Program Reference Manual* (EMTP Theory Book), prepared for Bonneville Power Administration, Department of Electrical Engineering, University of British Columbia, August 1986.
9. Bonner, A., Grebe, T., Gunther, E., Hopkins, L., Marz, M.B., Mahseredjian, J., Miller, N.W., Ortmeyer, T.H., Rajagopalan, V., Ranade, S.J., Ribeiro, P.F., Shperling, B.R., and Xu, W., Task Force on Harmonic Modeling and Simulation, Modeling and Simulation of the Propagation of Harmonics in Electric Power Systems, *IEEE Trans. Power Delivery*, 11(1), 452–465, 1996.
10. PSCAD/EMTDC User's Group home page, Transformer Inrush Current Simulation, http://pscad_µg.ee.umanitoba.ca/index.htm.
11. Electrotek Concepts, TOP—The Output Processor, http://www.pqsoft.com/top/.

9

Fundamentals of Power Losses in Harmonic Environments

9.1 Introduction

Estimation of harmonic-related losses in distribution systems entails the knowledge of harmonic sources, the characteristics of the elements involved in the propagation of the harmonic currents, and—what is most important and probably the most difficult to assess—the period during which harmonic currents are present in the system. Some applications involve well-identified periods of operation—for instance, the use of fluorescent lighting in commercial installations or the operation of electronic and digital equipment in business and other commercial facilities during working hours. Industrial facilities, however, are a special case because a variety of automated processes take place, many of them cyclic and often involving a mix of linear and nonlinear loads.

On the other hand, the operation of the electrical system, including operating voltage, substation, and service transformer configurations (primary and secondary connection types), and their leakage impedance, voltage regulation, and reactive power management practices also play an important role in the estimation of losses.

The utilization of harmonic cancellation schemes that eliminate higher-order harmonics is also relevant because losses are associated with the square of the current. Finally, the mobility of parallel resonant points as capacitor banks are turned on and off is also key for the estimation of harmonic-related losses.

This chapter presents a general description of the most relevant aspects to bear in mind when looking at losses in electrical systems related to harmonic currents.

9.2 Meaning of Harmonic-Related Losses

Increased rms values of current due to harmonic waveform distortion lead to increased heat dissipation in equipment and undesired fuse operation in capacitor banks. The resulting effect can affect life cycle due to accelerated aging of solid insulation in transformers, motors, and capacitor banks. It is usual to regard the dissipation of heat in electrical networks as the product, I^2R, evocative of electrical resistive losses. For the case of harmonic distortion, the total losses can be expressed as the summation of the individual losses, including every harmonic frequency. For a case involving fifth and seventh harmonics, the total losses in a 60 Hz system would be described as

$$I^2Z = I_{60\ Hz}^2\ Z_{60\ Hz} + I_{300\ Hz}^2\ Z_{300\ Hz} + I_{420\ Hz}^2\ Z_{420\ Hz} \tag{9.1}$$

Because most electrical equipment is specified based on 50/60 Hz, the addition of harmonic losses can limit the ability of the equipment to operate at rated parameters. Harmonic losses are related to the additional heat developed during the operation of nonlinear loads. Putting this in a simple perspective, harmonics losses can be regarded as the difference in heat dissipation between two parallel loads of the same size, one linear and the other nonlinear, when they are fed off from the same source.

Figure 9.1 illustrates this concept. I_h is the harmonic current that produces the additional losses. There is a linear load at the left of the figure and a nonlinear load to the right. The nonlinear load is a variable frequency drive, symbolized here, for simplicity, just as a thyristor. The output of the

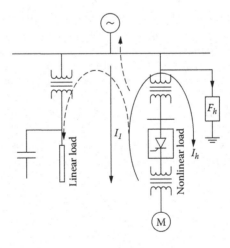

FIGURE 9.1
Harmonic current distribution from a nonlinear load affecting an adjacent facility.

drive is an AC voltage that, in the example, is stepped up to compensate the voltage drop on a long cable feeding a large load. As previously discussed in Chapter 2, variable frequency drives distort the source side voltage waveform due to the way in which they draw current in slices during the commutation process. The decomposition of the current waveform in Fourier series provides the spectrum of harmonic currents that can be used to calculate the combined contribution of every harmonic component to the total losses. Harmonic losses will generally show up in the form of copper and core losses.

Here it is important to remember that harmonic currents increase the rms or total effective load current as follows:

$$I_{RMS} = \sqrt{I_1^2 + \sum_{h=2}^{n} I_h^2} \tag{9.2}$$

If this current is used in the general equations that express ohmic losses, the result will describe the losses contributed by all individual harmonic currents. The example in Figure 9.1 is intended to illustrate in a simple way how harmonic loss dissipation may take place in the context of two similarly sized but different types of loads. Both of them draw similar power frequency currents that will produce identical heat dissipation in their feeding transformers. However, the increased rms current value due to harmonics from the nonlinear load will create added transformer losses in a different fashion, as follows.

If there are harmonic filters tuned to provide low-path impedances for all characteristic harmonics of the converter, the several branches of filter, F_h, will ideally absorb them all. In this way, the transformer connected to the nonlinear load is subject to additional losses.

If the filter is not there, harmonic currents will freely propagate upstream, finding an additional path toward any adjacent facility and toward the AC source, as indicated by the dotted lines in Figure 9.1. This will now create added losses on the two service transformers. In these circumstances, the two customer facilities are affected by additional losses arising from harmonic currents created at one of the customer premises.

Assuming the most likely scenario—that the branches of harmonic filter, F_h, are tuned to lower-order harmonics—the remaining spectral components, especially those of a higher order, may find a likely path toward the capacitor bank on the adjacent facility. Increased feeder and source impedance losses would need to be accounted for in the calculation of total increased losses. How exactly losses will distribute on the different components can only be determined through a detailed representation of the distribution system elements.

In any case, it is obvious that the transformer on the nonlinear load will be subject to the largest losses, regardless of the existence of the harmonic filter. Specially designed K-type transformers discussed in Chapter 4 are used in these applications; they are expected to dissipate the added energy loss adequately without increasing the transformer temperature beyond design limits. The question that arises here concerns which parameters play a role in the generation of harmonic losses. This is the topic of the next section.

9.3 Relevant Aspects of Losses in Power Apparatus and Distribution Systems

Under purely sinusoidal conditions, the calculation of losses in a power system is straightforward because it is based on conventional power flow studies that assume linear impedances throughout the system. The increasing waveform distortion in power systems due to the proliferation of nonlinear loads requires losses to be calculated using more suitable techniques. These involve time series in which voltage and current quantities are expressed comprising the most relevant frequency components other than the fundamental frequency of the system.

As a result, the active, reactive, and apparent power must be determined using the expressions presented in Chapter 1 as follows:

$$P = \frac{1}{T}\int_0^T p(t)dt = \sum_{h=1}^{\infty} V_h I_h \cos(\theta_h - \delta_h) = \sum_{h=1}^{\infty} P_h \qquad (9.3)$$

$$Q = \frac{1}{T}\int_0^T q(t)dt = \sum_{h=1}^{\infty} V_h I_h \sin(\theta_h - \delta_h) = \sum_{h=1}^{\infty} Q_h \qquad (9.4)$$

$$S = V_{\text{rms}} I_{\text{rms}} \qquad (9.5)$$

$$S^2 = P^2 + Q^2 + D^2 \qquad (9.6)$$

9.4 Harmonic Losses in Equipment

9.4.1 Resistive Elements

If we assume that a 1 Ω resistive element is the path for a fundamental current, $I_1 = 1$ A, containing additional third, fifth, and seventh harmonic levels whose amplitudes are inversely proportional to their harmonic order, the rms current can be calculated as follows:

$$I_{RMS} = \sqrt{\sum_{h=1}^{\infty} I_h^2} = I_{RMS} = \sqrt{1^2 + \left(\frac{1}{3}\right)^2 + \left(\frac{1}{5}\right)^2 + \left(\frac{1}{7}\right)^2} \approx 1.082 \qquad (9.7)$$

This small increase in current above 1 A will produce increased losses, ΔP, relative to the case in which current contained no harmonics. This yields

$$\Delta P = (I_{RMS}^2)(R) - (I_1)^2(R) = (I_{RMS}^2 - I_1^2)R$$
$$= [(1.082)^2 - (1)^2](1) = 0.1715 \text{ W} \qquad (9.8)$$

or

$$\frac{0.1715}{1}(100)$$

This is practically 17% above the case with no harmonics.

The corresponding THD_I is given by Equation (1.40) in Chapter 1 as follows:

$$THD_I = \frac{\sqrt{\sum_{h=2}^{\infty} I_h^2}}{I_1}$$

$$THD_I = \frac{\sqrt{\sum_{h=2}^{\infty}\left[\left(\frac{1}{3}\right)^2 + \left(\frac{1}{5}\right)^2 + \left(\frac{1}{7}\right)^2\right]}}{1} = 0.4141 = 41.4\% \qquad (9.9)$$

If the assumed current is 2 A with harmonic currents keeping the same proportion relative to the fundamental current, rms current and losses become

$$I_{rms} = \sqrt{2^2 + \left(\frac{2}{3}\right)^2 + \left(\frac{2}{5}\right)^2 + \left(\frac{2}{7}\right)^2} \approx 2.165 \text{ A}$$

and

$$\Delta P = \left(I_{rms}^2 - I_1^2\right) R = [(2.165)^2 - (2)^2](1) = 0.686 \text{ W}$$

or

$$\frac{0.686}{2}(100)$$

This is practically 34% higher than losses with no harmonics.

If we make P_1 and P_2 stand for the losses at 1 and 2 A, respectively, their ratio can be expressed as

$$\frac{P_2}{P_1} = \frac{2.165^2(1)}{1.082^2(1)} = 4$$

Thus, from this example, we observe that the loss increase is proportional to current even under distorted conditions, and that the total dissipated power on the resistor increases proportionally to the square of current; thus, if current doubles, losses quadruple.

The corresponding THD_I is the same as for the previous case because the harmonic currents assumed were increased in the same proportion as the fundamental current. From Equation (9.8),

$$THD_I = \frac{\sqrt{\sum_{h=2}^{\infty}\left[\left(\frac{2}{3}\right)^2 + \left(\frac{2}{5}\right)^2 + \left(\frac{2}{7}\right)^2\right]}}{2} = 0.4141 = 41.4\%$$

From Equations (9.7) and (9.9), we can determine the relationship between the rms current and the total harmonic distortion of the current:

$$I_{RMS} = \sqrt{\sum_{h=1}^{\infty} I_h^2} = \sqrt{I_1^2 + \sum_{h=2}^{\infty} I_h^2}$$

$$THD_I = \frac{\sqrt{\sum_{h=2}^{\infty} I_h^2}}{I_1}$$

$$THD_I^2 = \frac{\sum_{h=2}^{\infty} I_h^2}{I_1^2} = \frac{I_{RMS}^2 - I_1^2}{I_1^2} \tag{9.10}$$

$$(THD_I^2)(I_1^2) + I_1^2 = I_{RMS}^2$$

$$I_{RMS} = \sqrt{I_1^2(THD_I^2 + 1)}$$

$$I_{RMS} = I_1\sqrt{THD_I^2 + 1}$$

9.4.2 Transformers

Transformer losses have two components: copper and core losses. Copper losses occur in the windings and are a function of 50/60 Hz resistance; at increased frequencies, resistance is even increased due to skin effect. Several methods to estimate the additional heating expected from nonsinusoidal loads are discussed in the next sections.

9.4.2.1 Crest Factor

This is the simplest way to express the relation between the maximum and the effective value of a voltage signal and yields $\sqrt{2}$ for the case when the signal is a pure sinusoidal waveform. This ratio is exposed to changing under harmonic distortion of the voltage signal created by nonlinear loads. It was popular in 1988[1] to express the impact of harmonics on the voltage fed to computer equipment. The mathematical definition of crest factor is the peak magnitude of the current waveform divided by its true rms value:

$$\mathrm{CrestFactor} = \frac{V_{\mathrm{peak}}}{V_{\mathrm{RMS}}}$$
(9.11)

9.4.2.2 Harmonic Factor or Percent of Total Harmonic Distortion

This is the frequently cited total harmonic distortion (THD) factor expressed in Equation (9.9). THD factor measures the contribution of the additional rms harmonic current to the nominal rms fundamental current; however, similar to crest factor, it does not have a means to consider harmonic heating losses.

9.4.2.3 K Factor

The calculation of K factor considers the important effect that frequency has on transformer loss estimation. This factor is defined as the sum of the squares of the harmonic current in per unit (p.u.) times the square of the harmonic number. In the form of an equation:

$$K = \sum_{h=1}^{\infty} (I_h^2 * h^2)$$
(9.12)

Alternatively, it can also be expressed as

$$K = \frac{\sum_{h=1}^{\infty} h^2 I_h^2}{\sum_{h=1}^{\infty} I_h^2}$$
(9.13)

where h is the harmonic order and I_h is the harmonic current of order h expressed in p.u. of the fundamental frequency current.

As expressed by Equations (9.12) and (9.13), K factor takes into account the effect of I^2R, which relates to losses, for the fundamental frequency and for every harmonic current component. This is a relevant parameter on the assessment of premature aging of transformer windings because of dissipated heat in the form of copper and core losses due to spectral components of the current.

Because K factor takes into account the frequency parameter, it is regarded as the most precise method to estimate the harmonic content of nonlinear loads for the specification of distribution transformers. See Chapter 1 for an additional description of K-type transformers. K factor transformers are constructed so that the higher the K factor is, the higher the harmonic content that they can handle without additional heating. $K = 1$ would be a conventional transformer not fitted for working in harmonic environments at rated power.

Following Underwriters Laboratories listing of the K4 to K50 transformers aligned with ANSI C57.110-1986,[2] changes to transformer designs were made to minimize losses. Changes considered increasing the primary winding size to better tolerate the circulating triplen harmonics, getting a design with a lower flux density core and insulated parallel transposed secondary wiring conductors to reduce resistance involved in the skin effect heating. This looked promising to obtain transformer designs with improved thermal dissipation to minimize the additional losses.

K factor is then an index that determines the changes that conventional transformers must be subjected to so that they can adequately handle the additional iron and copper losses that will be imposed by harmonic currents, particularly when operating at rated power. This is a needed measure to avoid having to derate transformer nominal capacity when installed in harmonic environments.

9.5 Example of Determination of K Factor

Assume that the harmonic content observed at the point of common coupling (PCC) in an industrial facility is that shown in the first two columns of Table 9.1. Calculate the K factor using the expressions shown before.

Using the values obtained in Table 9.1, the K factor according to Equation (9.12) yields

$$K = \sum_{h=1}^{21} (I_h(p.u.))^2 \cdot h^2 = 9.21$$

TABLE 9.1

Harmonic Content at the PCC of an Industrial Facility

h	I_h	I_h^2	i_h^a	$i_h^2 h^2$	$I_h^2 h^2$
1	1	1	0.9099	0.8279	1
3	0.33	0.1089	0.3003	0.8116	0.9801
5	0.20	0.04	0.1819	0.8272	1
7	0.14	0.0196	0.1274	0.7953	0.9604
9	0.11	0.0121	0.1000	0.8100	0.9801
11	0.09	0.0081	0.0819	0.8116	0.9801
13	0.08	0.0064	0.0728	0.8956	1.0816
15	0.07	0.0049	0.0637	0.9129	1.1025
17	0.06	0.0036	0.0546	0.8615	1.0404
19	0.05	0.0025	0.0455	0.7474	0.9025
21	0.05	0.0025	0.0455	0.9124	1.1025
Σ		1.2086		9.21	11.13

a $i_h = \dfrac{I_h}{I_{\text{rms}}} = \dfrac{I_h}{(\Sigma I_h^2)^{1/2}}$.

Alternatively, using Equation (9.13), K factor results in

$$K = \frac{\displaystyle\sum_{h=1}^{21} h^2 I_h^2}{\displaystyle\sum_{h=1}^{21} I_h^2} = \frac{11.13}{1.2086} = 9.21$$

As observed, the calculated K factor results are the same using either of the two expressions described in Equations (9.12) and (9.13).

9.6 Rotating Machines

The difference between the synchronous actual speed of an induction motor (speed at which the magnetic field is rotating) and the actual rotor speed is known as slip frequency. The electromagnetic torque varies as a direct function of the slip. This means that a large electromagnetic torque would require of a large slip frequency, ω_{slip}.

Induction machine losses can be estimated as the difference between the power crossing the air gap through the rotor ($T\omega_{\text{sync}}$) and the power delivered through the rotor to the load ($T\omega_s$)[3]:

$$P_{\text{losses}} = T\omega_{\text{sync}} - T\omega_m = T\omega_{\text{slip}} \tag{9.14}$$

This suggests that a small slip will minimize induction machine rotor losses.

A study conducted by Fuchs et al.[4] assessed harmonic losses in the stator of an 800 W, 60 Hz, 4-pole, 1738 rmp, 2.35 A phase current and a 220 V induction motor having stator, rotor, and magnetization parameters as follows:

- $R_1^S = 7.0 \, \Omega$, $X_1^S = 8.0 \, \Omega$
- $R_1^R = 4.65 \, \Omega$, $R_1^R = 7.3 \, \Omega$
- $X_m = 110 \, \Omega$

The outcome of this study showed harmonic stator and rotor losses as a percentage of the total stator and rotor losses, as shown in Table 9.2 and Figure 9.2.

In summary, the findings of this study showed:

A significant effect from stator subharmonic losses with decreasing frequency

Rotor losses that are larger for negative-sequence harmonics

Increased rotor losses due to sub- and interharmonics for decreasing frequencies below the power system fundamental frequency

TABLE 9.2

Stator, Rotor, and Total Harmonic Losses and Additional Temperature Rise in Induction Machines

Total Harmonic Losses for 800 W Induction Motor as Percentage of Total Losses				
Harmonic Order	3	5	7	11	13
Stator	5.6	3.8	3.7	Less than 1%	Less than 1%
Rotor	17.2	9.1	6.7		

Measured Additional Temperature Rise of Stator 2 HP Motor Induction from Positive- and Negative-Sequence Harmonic Currents			
Harmonic Order	5	7	11	13
	3	2.1	1.8	1.2

Measured Additional Temperature Rise of Stator 2 HP Motor Induction from Positive-Sequence Harmonic Currents			
Harmonic Order	5	7	11	13
		3.8		1.8

Measured Additional Temperature Rise of Stator 2 HP Motor Induction from Negative-Sequence Harmonic Currents			
Harmonic Order	5	7	11	13
	5.2		2.9	

Source: Adapted from Fuchs, E.F. et al., *Trans. Power Delivery*, 19(4), 2004.

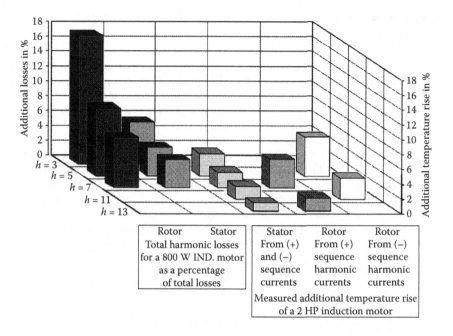

FIGURE 9.2
Harmonic losses and temperature increase in induction machines. (Adapted from Fuchs, E.F. et al., *Trans. Power Delivery*, 19(4), 2004.)

Similarly, increased temperatures were measured by Fuchs et al.[4] on a 2 HP squirrel cage, three-phase induction motor. The results, also summarized in Figure 9.2, revealed:

Slightly larger temperature rise due to negative-sequence harmonic voltages acting on the rotor

Rapidly increasing temperature rise from sub- and interharmonic voltage components as frequency decreased below the fundamental power system frequency

In the case of motors supplied from variable frequency drives, elevated levels of harmonics can be involved. This is particularly true when motors are operated at low frequencies. The power loss calculation should include in this case:

The power dissipated in the form of losses from the PCC down to the point at which the motor is supplied, including connection cables, transformers, and the variable frequency drive

The additional power that will be required if the motor is operated at frequencies above 50/60 Hz

References

1. Computer Business Equipment Manufacturers' Association (CBEMA), *Three-Phase Power Source Overloading Caused by Small Computers and Electronic Office Equipment*, ESC-3 Information Letter, November 1987.
2. ANSI/IEEE C57.110-1986, *Recommended Practice for Establishing Transformer Capability When Supplying Nonsinusodial Load Currents.*
3. Mohan, N., *Electric Drives: An Integrative Approach*, MNPERE, 2003. CEI/IEC 1000-2-1, *Electromagnetic Compatibility, Part 2: Environment, Section 1: Description of the Environment—Electromagnetic Environment for Low-Frequency Conducted Disturbances and Signaling in Public Power Supply Systems*, edition 1, 1990–05.
4. Fuchs, E.F., Roesler, D.J., and Masoum, M.A.S., Are Harmonic Recommendations according to IEEE and IEC Too Restrictive? *Trans. Power Delivery*, 19(4), 2004.

10

The Smart Grid Concept

10.1 Introduction

The smart grid, a concept so often mentioned, conveys the idea that power utilities are persistently working to achieve an electric power system that can smartly react and adjust to changes in the electrical demand or to disturbance conditions to continue delivering an interruptible and reliable power supply to all customers. In fact, not only power utility companies, but also governments, manufacturers, and private investors, are joining utilities worldwide in the effort to put together all that it takes in the integration, testing, and assessment of new devices meant to be part of an advanced electric system.

This is conceived as an adaptable system that will be able to manage itself, responding swiftly to operative changes and disturbances, including cyber security attacks, to make up a responsive and ultra-reliable power supply. This will require the use of advanced sensors, reactive power banks, energy storage devices, actuators, and communication systems to convert it into a smart grid.

It is important to understand that in modernizing the electric power system, substantial amounts of power will have to go through power conversion processes, as massive amounts of energy from distributed resources and stored energy devices are coupled to the electric grid.

As it was described in Chapter 2, rectification processes carry the generation of harmonic currents. However, the smart grid is likely to include processes other than rectification, including magnetic core saturation needed for some modern devices to operate, DC transmission, and flexible AC transmission system (FACTS) devices that use electronic switching to manage reactive power integration on the electrical network. Other devices that can potentially create harmonic distortion are static VAR compensators, solar and wind renewable energy integration in the electrical system, and energy storage devices, since they use different procedures of power conversion.

The increased use of electric vehicles (EVs) has also led to the idea that EVs can potentially become electricity providers to the power grid, acting just as small-scale distributed generators during the time they are parked, which in most cases comprises the greatest part of the time. This concept entails a two-way conversion of power so that electricity flow transitions between battery recharge cycles and provision of power to the grid. This concept is seen as a feasible initiative since it would essentially provide EV owners an attractive way to profit from selling electricity to the power supplier using the power storage capability of the vehicle. When this happens in large scale, without doubt the increasing levels of harmonic currents that will take place during peak battery recharge periods will create significant harmonic current increase with the potential to excite harmonic resonance conditions.

It is also worth mentioning that most of the cited applications may not create significant levels of harmonics per se. However, specific combinations of inductive and capacitive elements in the power system may excite resonance conditions at frequencies corresponding to one of the frequency components in the spectrum of the harmonic currents created in the process and magnify total harmonic distortion (THD) levels.

So, the modernization of the electric power grid coupled with the growing demand will invariably introduce some increased level of harmonic currents. To properly cope with this trend and the prospective proliferation of high-power frequency components above tolerable limits, it is important to review the present government and private investments in the smart grid concept for the coming years.

As of September 2013, the smart grid government in the United States had afforded $3.4 billion, and private funds in the amount of $4.5 billion had been applied to support electric transmission and distribution systems, advanced metering infrastructure (AMI), and customer systems.[1] This makes up an impressive funding of nearly $8 billion to help the initial integration, assessment, and deployment of new smart grid technologies in order to meet the nation's energy and environmental priorities.[1] The report describes this as a small down payment for the hundreds of billions of dollars that will be required in the next decades to fully modernize the electric power grid in the nation. A different source[2] adds that the estimations for fully renovating the electric systems in the United States will fall in the range of $338 billion to $446 billion over the next 20 years, with around 70% of the costs allocated for distribution and an estimated 60 million smart meters in use by 2020.

A forecast mentioned in reference 2 estimates that the cumulative investment in the European Union for the period 2010–2020 will reach €56.5 billion, and that an investment of €1.5 trillion will be required in the period 2007–2030 to modernize generation, transmission, and distribution of the electrical grid. The number of smart meters expected to be deployed by 2020 in Europe is around 240 million. Figure 10.1 depicts the location of budgets and project categories in the European Union illustrated in reference 2.

Smart grid investments elsewhere are no doubt being undertaken by utilities and investors. As with any other new technology, an improved understanding of the benefits and limitations of smart grid devices is a commitment that every engineering professional involved in the planning, design, integration, testing, and assessment phases should have. The particular aspects addressed in this book are only those related to waveform

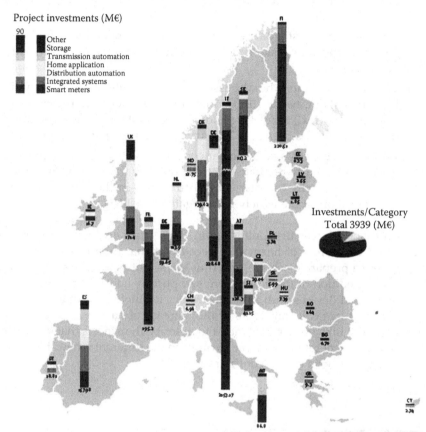

* This figure does not include the total budget of the Swedish smart meter programme (estimated budget € 1.5 billion), as not enough details were made available at this stage.

FIGURE 10.1 (See color insert.)
European investments in modernizing the electrical industry. (Adapted from Smart Grid Projects in Europe: Lessons Learned and Current Development; From Giordano, V. et al., Joint Research Centre for Energy Conference Records, Institute for Energy, EUR 24856 EN, European Union, 2011.)

distortion, how they fit within the established or recommended levels in industry standards, and how they can possibly influence operational aspects of the power system.

Other questions that deserve discussion and consideration are

1. Are new renewable resources pointed to pose a concern on wind and solar power penetration regarding their potential to raise prevailing harmonic distortion levels?

2. Shall we be revisiting present harmonic distortion levels so that they can accommodate the increasing penetration of renewable

power resources as they are regulated by power system demand control schemes?

3. Will the industry standards relax their recommended limits for harmonic distortion considering the intermittency of wind and solar power sources?

4. Will new regulations be discussed in the Institute of Electrical and Electronics Engineers (IEEE) and International Electrotechnical Commission (IEC) regulating bodies to account for the potential use of electric vehicles-to-grid (V2G) as distributed generation resources?

5. Will the increasing use of power converters with solid-state power electronics used to regulate reactive power, make DC transmission possible, and achieve other control functions in the context of the smart grid offset the widely promulgated benefits?

When this book was first published in 2006, smart grid devices were not mentioned as sources of harmonics because at that time it was a concept undergoing a preparatory stage, and thus it was not as mature as it is today. This revised edition of the book intends to portray how and to what extent new smart grid devices with solid electronics technology or other elaborate operation features can contribute to the increase of harmonic levels in the electrical grid. Likewise, it is intended to describe the aspects that determine the amount of generated harmonic currents in the different renewable power devices or smart grid device configurations.

10.2 Photovoltaic Power Generator

Solar power has been historically harnessed for different purposes, including water heaters, cooking, and solar concentrators to light fires, among others. However, solar energy did not start to be conceived as a potential source for electricity on a larger scale until the late 1950s, when photovoltaic cells more resistant to radiation were developed for cellar cells used in space applications.

NASA and the U.S. government through the Department of Energy and the National Renewable Energy Laboratory played a leading role in this effort, followed by undertakings from California power utilities. Just recently, the largest thermal solar plant was commissioned to supply PG&E and SCE in California around 377 MW by the end of 2013, representing a 100% increase in the amount of solar energy available in the United States. This is part of California's goal to provide around 30% of electricity from solar by 2020.[3] Similar efforts in India, Spain, France, Germany, Australia, Japan, Kenya, and the UAE are underway.

It is estimated that around 150 square miles in a desert area has the potential to generate around 20,000 MW from solar power, which would potentially power 20 million houses in the United States.[4]

The great support that this titanic renewable energy resource represents in a worldwide scale is evidence that we are witnessing the dawn of industry embracing the costly investments and extensive integration and deployment of massive solar cell arrangements to harnessing this abundant resource as a means to provide power to industry and to large sectors of the population. We just need to ensure that the energy conversion processes required to integrate the produced energy to the electric grid will be properly planned and designed so that waveform distortion levels in excess of acceptable industry recommended limits do not become an issue.

10.3 Harnessing the Wind

Wind power is a subproduct of solar energy. Wind is produced after solar energy heats the earth and warm air ascends, producing neighboring colder air to occupy the displaced ascending air.

The wind energy industry is booming. Globally, generation more than quadrupled between 2000 and 2006. At the end of 2013, global capacity was more than 300,000 megawatts. In the United States, a single megawatt is enough electricity to power about 250 homes. According to reference 5, worldwide, the People's Republic of China (PRC) has the lead in cumulative wind power capacity, followed by the United States, Germany, Spain, India, and the UK. Regarding installed capacity added in 2013, PRC, Germany, the UK, India, Canada, and the United States were among the top six, as illustrated in Figure 10.2.

Industry experts predict that if this pace of growth continues, by 2050 the answer to one-third of the world's electricity needs will be found blowing in the wind.

10.4 FACTS Technology Concept and Its Extended Adoption in Distribution Systems

Flexible AC transmission systems (FACTS) are widely recognized as a sound concept to help power systems operate in such a way that reactive power is properly integrated into the power system so that losses in lines, transformers, and other equipment are kept low, boosting life

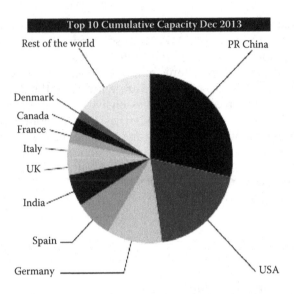

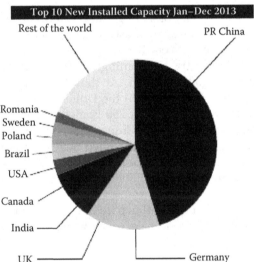

FIGURE 10.2 (See color insert.)
Worldwide wind power cumulative capacities. (Adapted from American Recovery and Reinvestment Act of 2009, *Smart Grid Investment Grant Program*, Progress Report II, U.S. Department of Energy, Electric Delivery and Energy Reliability, October 2013.)

expectancy of equipment, improving voltage profile, and overall, offering an enhanced power quality. This is traditionally achieved through the use of capacitor banks switched on and off as needed, utilizing power electronic technology.

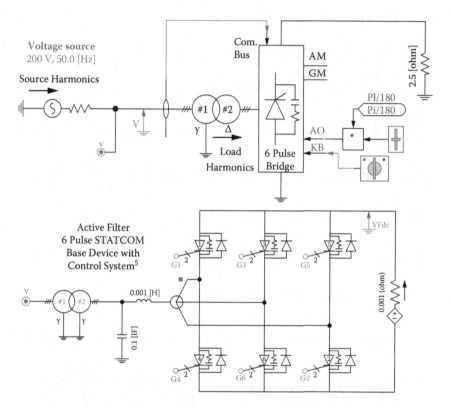

COLOR FIGURE 6.25
PSCAD model of a STATCOM-based active filter. (Adapted from Fujita, H., and Akagi, H., A Practical Approach to Harmonic Compensation in Power Systems—Series Connection of Passive and Active Filters, *IEEE Trans. Ind. Appl.*, 17(6), 1020–1025, 1991.)

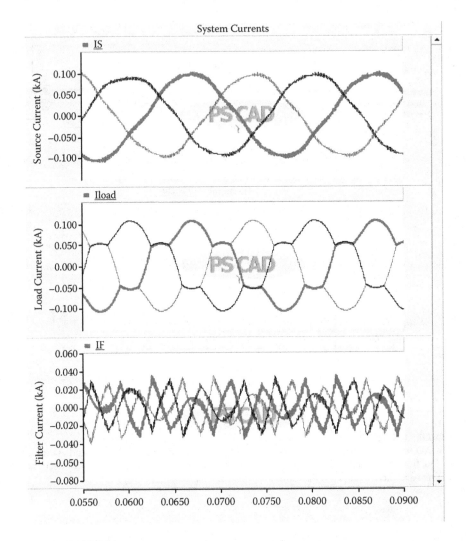

COLOR FIGURE 6.26
Source, load, and filter currents.

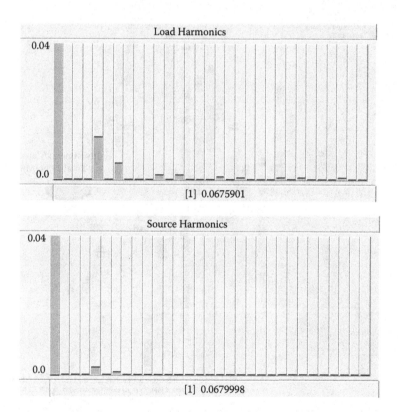

COLOR FIGURE 6.27
Harmonic spectra of the load and source currents, respectively, from the example shown with the active filter installed.

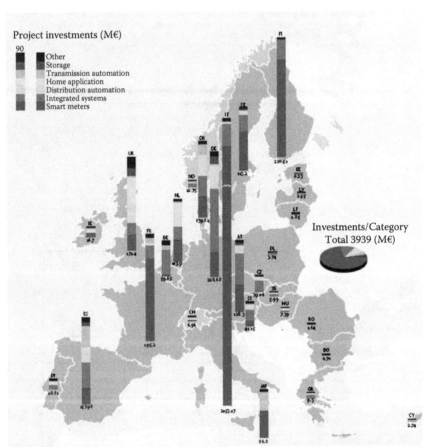

* This figure does not include the total budget of the Swedish smart meter programme (estimated budget € 1.5 billion), as not enough details were made available at this stage.

COLOR FIGURE 10.1
European investments in modernizing the electrical industry. (Adapted from Smart Grid projects in Europe: Lessons learned and current development; From Giordano, V. et al., Joint Research Centre for Energy Conference Records, Institute for Energy, EUR 24856 EN, European Union, 2011.)

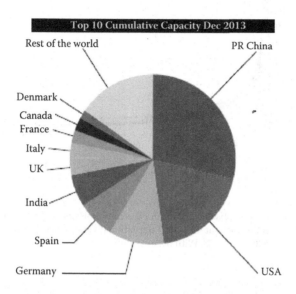

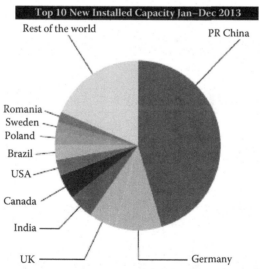

COLOR FIGURE 10.2
Worldwide wind power cumulative capacities. (Adapted from American Recovery and Reinvestment Act of 2009, *Smart Grid Investment Grant Program*, Progress Report II, U.S. Department of Energy, Electric Delivery and Energy Reliability, October 2013.)

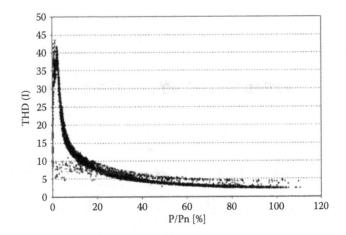

COLOR FIGURE 11.2
Total harmonic distortion of a photovoltaic inverter ($PDC/AC = 5$ kW) for different loading conditions. (Adapted from Schlabbach, J., Harmonic Current Emission of Photovoltaic Installations under System Conditions, presented at 5th International Conference on European Electricity Market 2008 (EEM 2008), May 2008. With permission.)

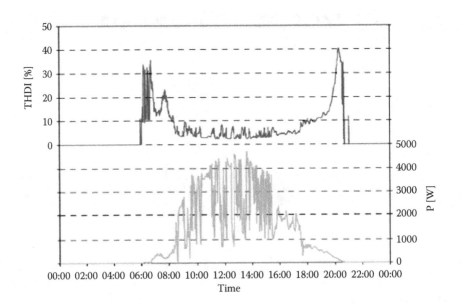

COLOR FIGURE 11.3
THD_I (upper curve) and generated power (lower curve) of the photovoltaic inverter ($PDC/AC = 5$ kW). (Adapted from Schlabbach, J., Harmonic Current Emission of Photovoltaic Installations under System Conditions, presented at 5th International Conference on European Electricity Market 2008 (EEM 2008), May 2008. With permission.)

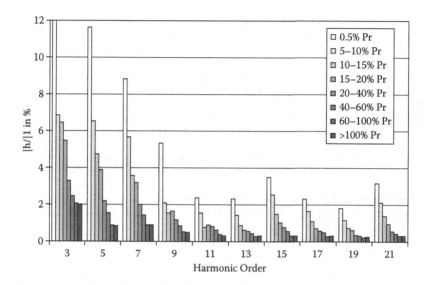

COLOR FIGURE 11.6
Current harmonics of the photovoltaic inverter (PDC/AC = 5 kW) for different loading conditions. (From Schlabbach, J., Harmonic Current Emission of Photovoltaic Installations under System Conditions, presented at the 5th International Conference on European Electricity Market 2008 (EEM 2008), May 2008.)

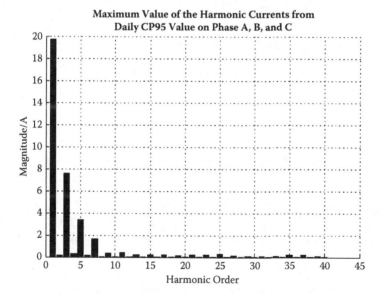

COLOR FIGURE 11.7
Current harmonics from measurements in a photovoltaic inverter. (From Ahmed, A.L., Harmonic Impact of Photovoltaic Inverter Systems on Low and Medium Voltage Distribution Systems, MEng thesis, School of Electrical, Computer, and Telecommunications Engineering, University of Wollongong, 2006.)

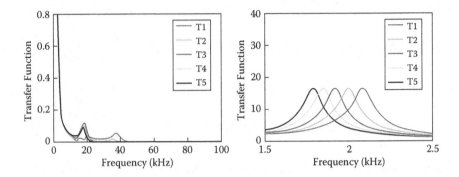

COLOR FIGURE 11.8
Individual transfer function for several turbines. (Adapted from Bollen, M., and Yang, K., *Elforsk Rapport*, 12, 51, 2012. With permission.)

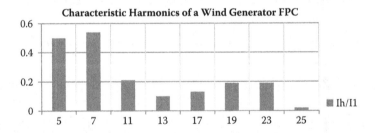

COLOR FIGURE 11.9
Harmonic spectrum of a wind power generator. (Adapted from Romero, A.A. et al., Possibilistic Harmonic Load-Flow for Electric Power Systems with Wind Farms, SICEL, Colombia, December 2013. With permission.)

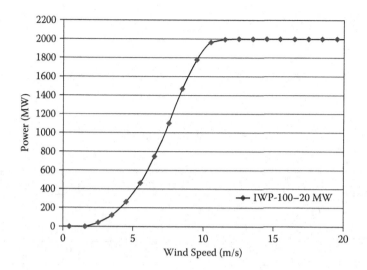

COLOR FIGURE 11.10
Power as a function of wind for a 2 MW full-power converter wind generator. (Adapted from Romero, A.A. et al., Possibilistic Harmonic Load-Flow for Electric Power Systems with Wind Farms, SICEL, Colombia, December 2013. With permission.)

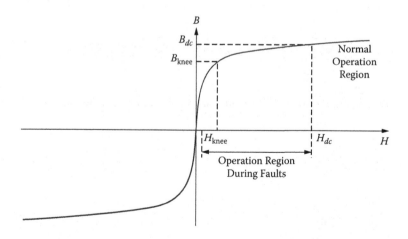

COLOR FIGURE 12.3
Simplified representation of a saturated-core SCFCL.

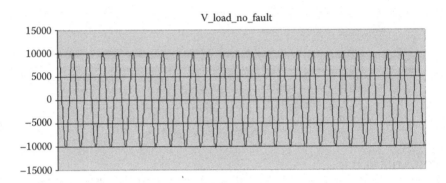

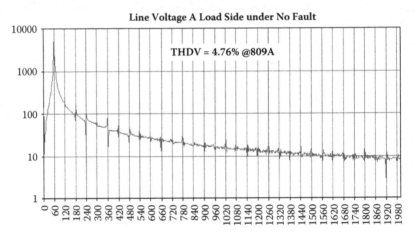

COLOR FIGURE 12.5
Line voltage and calculated THD for a saturated-core fault current limiter.

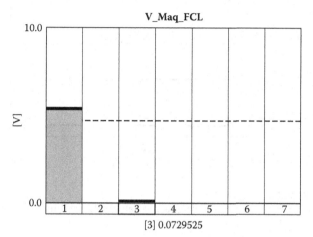

[3] 0.0729525

Harmonic Order

(a) Light load conditons

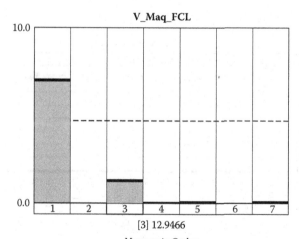

[3] 12.9466

Harmonic Order

(b) Full load conditons

COLOR FIGURE 12.6
Harmonic spectra of the calculated voltage across saturated-core fault current limiter for light and full load conditions.

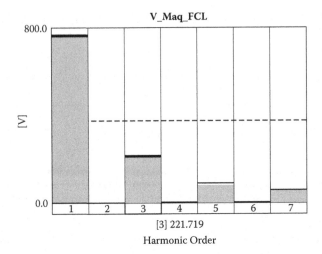

COLOR FIGURE 12.7
Harmonic spectra of the calculated voltage across saturated-core fault current limiter under a loss of DC bias implausible event.

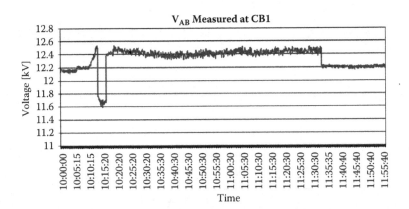

COLOR FIGURE 12.8
Line-to-line voltage in the SCE example at the time of the loss of DC bias in the SCFCL. (Adapted from Christopher, R.C. et al., Resonance of a Distribution Feeder with a Saturable Core Fault Current Limiter, presented at 2010 IEEE T&D Conference, New Orleans, April 19–22, 2010. With permission.)

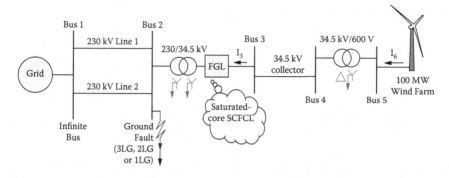

COLOR FIGURE 12.9
Saturated-core SCFCL as a means to reduced transient torque during faults. (Adapted from Muljadi, E. et al., Wind Power Plant Enhancement with a Fault Current Limiter, presented at Proceedings of 2011 IEEE Power and Energy Society General Meeting, Detroit, Michigan, July 24–29, 2011. With permission.)

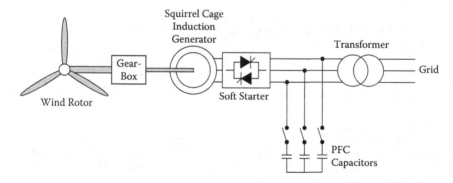

COLOR FIGURE 12.10
Wind farm with type I wind turbine induction generator. (Adapted from Muljadi, E. et al., Wind Power Plant Enhancement with a Fault Current Limiter, presented at Proceedings of 2011 IEEE Power and Energy Society General Meeting, Detroit, Michigan, July 24–29, 2011. With permission.)

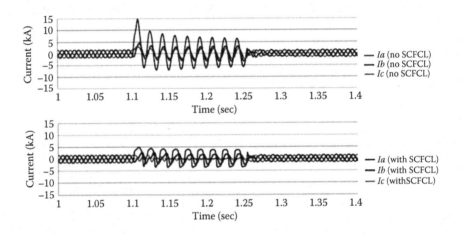

COLOR FIGURE 12.11
Single-phase-to-ground fault at the substation transformer. (Adapted from Muljadi, E. et al., Wind Power Plant Enhancement with a Fault Current Limiter, presented at Proceedings of 2011 IEEE Power and Energy Society General Meeting, Detroit, Michigan, July 24–29, 2011. With permission.)

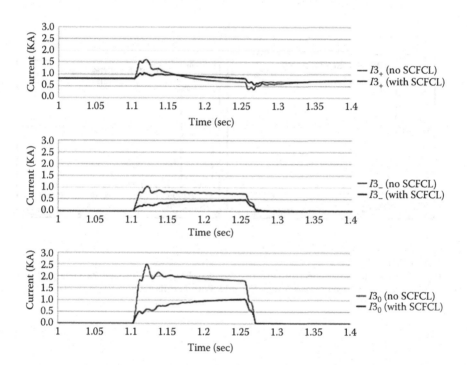

COLOR FIGURE 12.12
Sequence currents with and without SCFCL in the circuit. (Adapted from Muljadi, E. et al., Wind Power Plant Enhancement with a Fault Current Limiter, presented at Proceedings of 2011 IEEE Power and Energy Society General Meeting, Detroit, Michigan, July 24–29, 2011. With permission.)

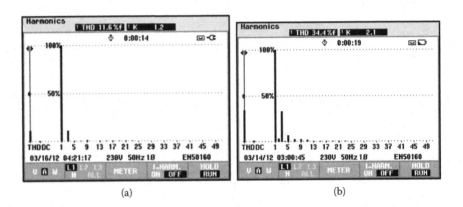

COLOR FIGURE 12.13
Typical current harmonic spectrum in (a) modern EV charger and (b) a golf cart charger. (From Wanic, M.Z.C. et al., *Engineering*, 5, 215–220, 2013. With permission.)

This all is carried out in transmission power substations. The consequence of this is that the needed equipment is often oversized to be able to cope with the volumes of energy that have to be handled.

The most common FACTS devices and their planned functions are depicted in Table 10.1. It is important to realize that most of the intended outcome is targeted at providing voltage support, since preserving the quality of voltage supply is one of the premises behind FACTS. Reactive power compensation by itself is not fully substantiated until it is expressed in the context of the benefits obtained in keeping voltage and delivered power within power quality targets.

All of the above devices are fitted with power electronic switching devices that may involve the following:

- Gate turn-off thyristor (GTO)
- Gate commutated thyristor (GCT)
- Thyristor-controlled series compensator (TCSC)
- Thyristor-controlled phase shift transformer (TCPST)
- Integrated gate commutated thyristor (IGCT)
- Insulated gate bipolar transistor (IGBT)

TABLE 10.1

FACTS Devices and Power System Parameters They Control

FACTS Device	Function
Static VAR compensator (SVC)	Controls voltage by providing fast-acting reactive power on high-voltage electricity transmission networks
Static synchronous compensator (STATCOM)	Regulates voltage by providing reactive power in response to voltage transients, enhances transient stability, improves damping of interarea oscillations
Unified power flow controller (UPFC)	Controls reactive power, voltage
Convertible series compensator (CSC)	Manages reactive power, voltage imbalance, current and voltage harmonic compensation
Interphase power flow controller (IPFC)	Manages reactive power, voltage, angle
Static synchronous series controller (SSSC)	Controls phase, voltage
Thyristor-controlled series compensator (TCSC)	Modifies impedance
Thyristor-controlled phase shifting transformer (TCPST)	Adjusts angle
Superconducting magnetic energy storage (SMES)	Supports voltage and power

A description of each of these technologies is beyond the scope of this book. The interested reader may find reference 6 an excellent source for an insightful tour into the specifics of the listed FACTS devices.

More often than not, the implied high cost of FACTS devices is prohibitive for power utilities. This has led to the idea of extending this concept to distribution class substations, where the size of the equipment and shorter lengths of lines, as well as diminished levels of active and reactive power involved, would make the objective look more plausible. Therefore, a significant increment in power electronics devices as part of the potential implementation of FACTS-type devices, and hence an increase in harmonic levels in the electric power distribution network, should be expected.

Figure 10.3 shows a typical installation of FACTS on a wind farm in Australia, including a low-voltage modular IGBT inverter-based centralized STATCOM, the wind farm wind turbine generators (WTGs), and the power factor capacitors all connected to the medium-voltage substation bus.[7]

Whatever the case may be, the implication of using electronic switching in these applications is that the resultant harmonic currents produced by solid-state switching in the rectification and inversion processes are likely to lead to inconvenient levels of individual and total harmonic distortion levels.

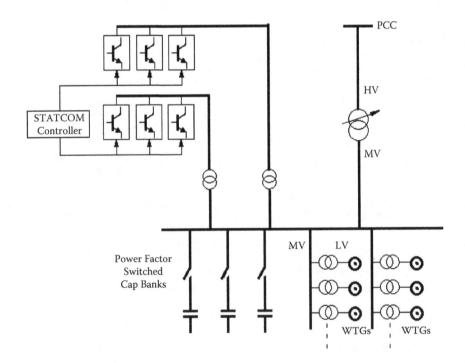

FIGURE 10.3
Typical connection of a STATCOM in a wind farm.

10.5 High-Voltage Direct Current (HVDC) Transmission

Currently, we notice the need to transmit increased levels of electric power. Frequently this involves long distances, and this leads to the necessity to assess whether AC or DC transmission is the best choice to keep the power losses low and preserve the transmission system operational and with power balance within desired targets. Reliability, operation, and economics are significant aspects to assess as well.

Figure 10.4 shows a simplified representation of the classical HVDC concept. Obviously, a DC power transmission link requires a converting station at both ends of the line.

Converter stations usually include converter transformers, thyristor valves, smoothing reactors, AC filters, and DC filters. The so-called light HVDC version[8] replaces the line commutated thyristor valves with self-commutated IGBT valves. Among other advantages, it further offers independent power transfer and voltage control, low power operation, power reversal, reduced power losses and increased transfer capacity.

With transfer capability unaffected by distance, lower line losses, and it being more environmentally friendly, HVDC is a natural strong competitor for AC transmission. Within their respective boundaries, classical HVDC offers transmission capacities in the ±800 kV 6000 MW range, whereas light HVDC is targeted to capacities in the ±320 kV 1200 MW limits.[8]

Classic HVDC has seen substantial progress from the originally employed mercury arc valves to line commutated thyristor valves. Although classic HVDC requires substantial reactive power compensation and low-order harmonic filters, it can accomplish fast active power control and reduced converter losses. As a compact version, the light HVDC concept uses self-commutated insulated gate bipolar transistor (IGBT) valves, it does not

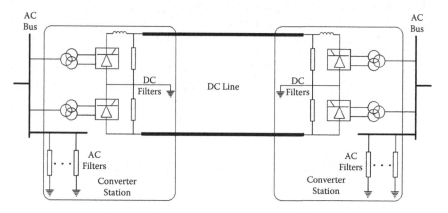

FIGURE 10.4
Simplified representation of the classical HVDC line system.

require reactive compensation, and it provides a fast active and reactive power control, while yielding converter losses two to three times higher than the converter losses in the classical technology. Regarding harmonics, in the light HVDC they are generated at switching frequencies in the 2 kHz region with almost no low-order harmonics as in the thyristor control drive.

With an increased power transfer capability as a function of distance, smaller losses, and environmental responsiveness, along with interarea oscillation damping, quick power flow controls, and a convenient dynamic voltage support, the light option HVDC may be heading to potential growth in HVDC applications.

References

1. American Recovery and Reinvestment Act of 2009, *Smart Grid Investment Grant Program*, Progress Report II, U.S. Department of Energy, Electric Delivery and Energy Reliability, October 2013.
2. Giordano, V., Gangale, F., Fulli, G., Sánchez-Jiménez, M., Onyeji, I., Colta, A., Papaioannou, I., Mengolini, A., Alecu, C., Ojala, T., and Mashio, I., *Smart Grid Projects in Europe: Lessons Learned and Current Developments*, EUR 24856 EN, Joint Research Centre for Energy Conference Records, Institute for Energy, European Union, 2011.
3. Tweed, K., World's Largest Solar Thermal Plant Syncs to the Grid, *IEEE Spectrum*, September 26, 2013.
4. Stafford, B., *The History of Solar*, U.S. Department of Energy, Energy Efficiency and Renewable Energy.
5. GWEC, *The Global Status of Wind Power in 2013*, Global Wind Report: Annual Market Update 2013.
6. Hingorani, G.N., and Gyugyi, L., *Understanding FACTS: Concepts and Technology of Flexible AC Transmission Systems*, New York, IEEE Press, 2000.
7. Mutik, P.K., Brown, R.W., Chamers, C., Gee, G.G., Haddow, J., and Pahalawaththa, N., FACTS for Enabling Wind Power Generation, in *CIGRE General Meeting, CIGRE 2010*, Session Paper B4-205-2010.
8. ABB, *It Is Time to Connect—Technical Description of HVDC Light Technology*, 2008.

11

Harmonics in the Present Smart Grid Setting

11.1 Introduction

There are abounding sources of harmonics in the smart grid concept. For the greatest part this is due to the extensive use of high-frequency power converters in switchable capacitors and reactors applied in reactive power management, coupling of wind and solar renewable resources to the power grid, high-voltage direct current (HVDC) transmission, and vehicle-to-grid interaction, among others, Certainly, for the most part, harmonics created in many of these applications may be inconsequential, but the occurrence of harmonic resonance is always present, for which a harmonic distortion assessment considering all power frequency components along with the reactive elements of the power system is important to carry out. This will allow for the implementation of remedial measures to control the harmonic current production and their potential amplification under harmonic resonance conditions within recommended limits.

11.2 Photovoltaic Power Converters

11.2.1 Main Operation Aspects

Solar power is increasingly being harnessed as an immense renewable resource by the power industry. Invariably, solar panels generate DC power, which needs consecutively to be transformed into AC power to connect to the AC grid or to local AC loads. This process is regularly called inversion, and it generates harmonics in the grid.

Figure 11.1 shows a block diagram of a typical solar inverter, including solar panels and all that is required to connect to the distribution line. Power from the photovoltaic modules in the form of DC is inverted, normally through a phase wave modulation (PWM) inverter, into AC power and subsequently transformed to the required voltage through a step-up transformer. The filters shown are needed to clean the waveform from the high-frequency components created in the electronic switching process. Two principal sources of

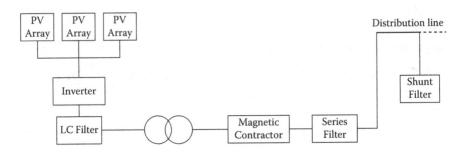

FIGURE 11.1
Block diagram of a typical photovoltaic system.

high-frequency noise can be identified. One is the modulation frequency of the PWM inverter, typically in the 2 to 20 kHz region. These high-frequency fluctuations are attenuated down the way by the LC filter and the transformer. The other source of high-frequency components is the switching transients produced during the operation of the power electronics switching devices, commonly the insulated gate bipolar transistors (IGBTs). Switching transient frequencies can well be as high as 100 MHz. The series and the shunt filters are provided to attenuate the frequency components caused by these switching transients, as well as by harmonics from other subsystems, like switched mode power supplies and other inverter control schemes.

The total harmonic distortion in the process occurs when the generated power is low, and it shows a lower level when the generated power is high, as depicted in Figures 11.2 and 11.3.

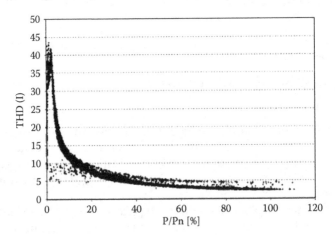

FIGURE 11.2 (See color insert.)
Total harmonic distortion of a photovoltaic inverter ($PDC/AC = 5$ kW) for different loading conditions. (Adapted from Schlabbach, J., Harmonic Current Emission of Photovoltaic Installations under System Conditions, presented at the 5th International Conference on European Electricity Market 2008 (EEM 2008), May 2008. With permission.)

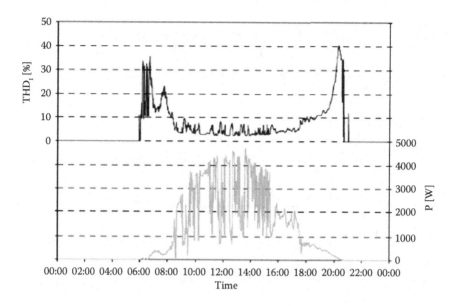

FIGURE 11.3 (See color insert.)

THD_I (upper curve) and generated power (lower curve) of the photovoltaic inverter ($PDC/AC =$ 5 kW). (Adapted from Schlabbach, J., Harmonic Current Emission of Photovoltaic Installations under System Conditions, presented at the 5th International Conference on European Electricity Market 2008 (EEM 2008), May 2008. With permission.)

11.2.2 Harmonic Generation in Photovoltaic Converters

The output waveform from a PWM converter is shaped from the comparison of a reference signal (sinusoidal waveform) and a carrier waveform (a triangular waveform) of frequency called modulation frequency, as depicted in Figure 11.4. When the reference signal is larger than the carrier

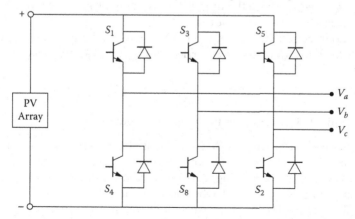

FIGURE 11.4

Typical PWM inverter.

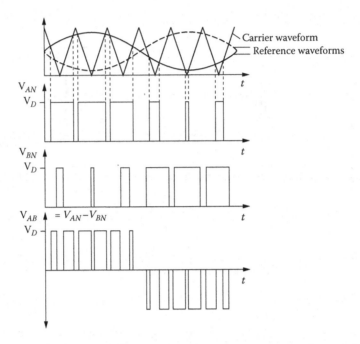

FIGURE 11.5
Output voltage creation in a PWM inverter.

waveform, the upper IGBT in Figure 11.4 is switched on, with the lower IGBT staying off, and positive DC voltage is applied to the inverter output phase (A). When the reference signal is inferior to the carrier waveform, the lower IGBT is switched on (with the upper IGBT staying off) and negative DC voltage is applied to the inverter output.

The fundamental principle of the PWM inverter is graphically depicted in Figure 11.5. Sinusoidal reference, carrier, and V_{AN}, V_{BN}, and V_{AB} waveforms are illustrated.

11.2.3 Typical Harmonics in Photovoltaic Converter

Harmonics generation in a PWM inverter, as stated earlier, is also a function of loading. Figure 11.6 illustrates the harmonic spectra in a 5 kW converter for varying degrees of loading. The preponderant odd harmonic presence in the spectrum is confirmed when looking at the frequency components observed in Figure 11.7 from measurements in an Australian work reported in reference 2.

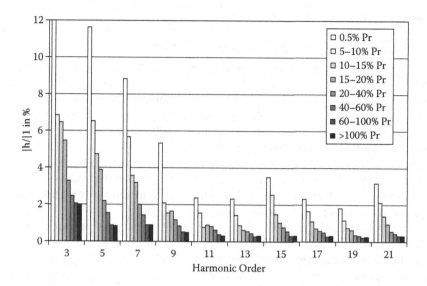

FIGURE 11.6 (See color insert.)
Current harmonics of the photovoltaic inverter (*PDC/AC* = 5 kW) for different loading conditions. (From Schlabbach, J., Harmonic Current Emission of Photovoltaic Installations under System Conditions, presented at the 5th International Conference on European Electricity Market 2008 (EEM 2008), May 2008.)

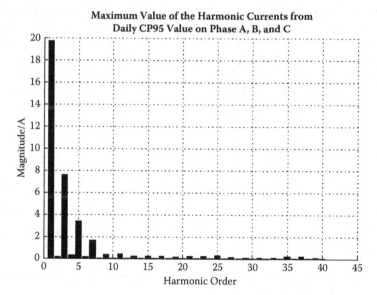

FIGURE 11.7 (See color insert.)
Current harmonics from measurements in a photovoltaic inverter. (From Ahmed, A.L., Harmonic Impact of Photovoltaic Inverter Systems on Low and Medium Voltage Distribution Systems, MEng thesis, School of Electrical, Computer, and Telecommunications Engineering, University of Wollongong, 2006.)

11.3 Conventional Wind Power Converters

Wind power, discussed in this section, relates to those applications where a power converter/coupling transformer is used as an interface between the wind turbine and the collector. Particular situations involving a connection of the wind turbine to the collector via a coupling transformer do not create harmonics, unless the transformer is operated under saturation, and are therefore outside of this discussion.

Another aspect that has been largely discussed in many references relates to the significantly different nature of harmonic distortion produced in wind parks relative to background distortion existing in the power grid prior to the interconnection of the wind park.

On the one hand, harmonic distortion in wind parks is produced within the turbines, but primarily by the solid electronics in the power converters. On the other hand, background distortion can be produced by any individual or combined contribution from nonlinear loads or from harmonic amplification produced by resonant conditions.

Harmonic distortion at the collector level is further influenced by some form of harmonic component cancellation from different turbines connected to the grid.[3] The referenced work assesses harmonic distortion by way of a transfer function described as the ratio between individual harmonic levels of specific turbines and those on the collector at the connection point, with all other emissions neglected. The individual transfer function showed amplification levels around the 1 and 2.2 kHz frequency ranges, with an emission from the grid being around 20 times the emission from individual turbines due to resonances in the connection grid, and falling drastically at higher frequencies, as illustrated in Figure 11.8. T1 through T6 refer to

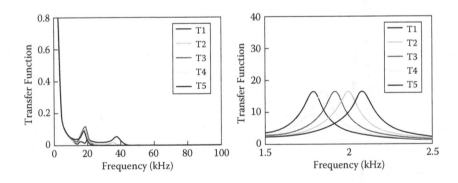

FIGURE 11.8 (See color insert.)
Individual transfer function for several turbines. (Adapted from Bollen, M., and Yang, K., *Elforsk Rapport*, 12, 51, 2012. With permission.)

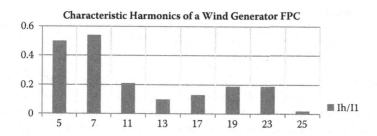

FIGURE 11.9 (See color insert.)
Harmonic spectrum of a wind power generator. (Adapted from Romero, A.A. et al., Possibilistic Harmonic Load-Flow for Electric Power Systems with Wind Farms, SICEL, Colombia, December 2013. With permission.)

different wind farm turbines. All of them were around 2 MW, and all connected to the grid through power converters.

11.3.1 Typical Harmonics in Wind Power Converters

Figure 11.9 describes the characteristic harmonics of a wind generator full-power converter, and Figure 11.10 depicts the power as a function of wind speed in a 2 MW converter reported in reference 4.

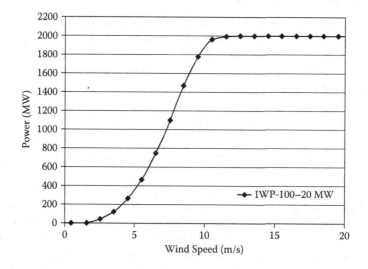

FIGURE 11.10 (See color insert.)
Power as a function of wind for a 2 MW full-power converter wind generator. (Adapted from Romero, A.A. et al., Possibilistic Harmonic Load-Flow for Electric Power Systems with Wind Farms, SICEL, Colombia, December 2013. With permission.)

11.4 Power Electronics Harmonics Inherent in FACTS Technology

One of the most common flexible AC transmission system (FACTS) technology devices is the switched capacitor bank, which is used to provide the required reactive compensation. This entails knowledge of the existing levels of active and reactive power in the electrical system. Essentially, varying amounts of reactive power are incorporated or removed from the grid through electronic switching of individual capacitor bank units.

This is associated with the variations of demand during the day, which can show particular patterns depending on a number of factors.

What all this means is that varying amounts of capacitor bank power are expected during a daily cycle, with peaks and valleys occurring in time very much depending on season of the year and day of the week, type of customers, and geographical area, which determines the climate and power demand patterns. The amount of capacitive power also influences the likelihood of having resonant conditions in the power system aligning at different times of the day with natural frequency components from the harmonic-rich spectra from power converters. This can potentially produce an amplification of THD_V/THD_I levels.

11.4.1 Most Common Power Frequency Components in the FACTS Technology

For a balanced AC system FACTS converters generate characteristic harmonics that can be directly found from the switching function, which is determined from the number of pulses in the converter. Consequently, the AC side current distortion (h_ac) occurs at multiples of the PWM switching frequency (fs), expressed in terms of harmonic order as

$$h\_dc = (jhs \pm k), \quad j, k = 1, 2, 3, \ldots \tag{11.1}$$

For unbalanced systems, FACTS converters may produce noncharacteristic and even harmonics.

FACTS technology has a rich variety of possibilities, as described in condensed form in Figure 11.11, and is fundamentally aimed at providing flexible control of transmission systems. The main applications of FACTS reside in achieving reactive power control. Potentially, FACTS technology can help defer costly transmission line upgrades or supplant obsolescent generation. However, the use of electronic switching in any of these devices, even if they are provided with harmonic filters, conveys the possibility of resonance involving the system impedance. This may be controversial given that FACTS devices are also used to mitigate harmonic levels in the transmission

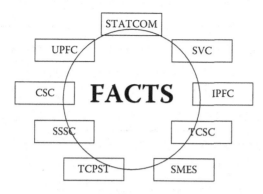

FIGURE 11.11
Flexible AC transmission technology devices.

systems. However, resonant conditions always have to be investigated at the specific levels of reactive capacitive compensation provided by these devices to look out for the possibility of harmonic resonance.

Some of these devices are shunt connected to the transmission system to provide fast voltage and reactive power control or damping of oscillations of power. The static volt ampere reactive (VAR) compensator (SVC) and the STATCOM[5] are probably the most commonly shunt-connected FACTS devices since they can enable the large-scale integration of wind power, which requires voltage control.

In long transmission lines requiring series compensation to reduce transmission angle and improve stability, thyristor-controlled series compensation (TCSC) is commonly used to control power flow in parallel lines or provide power oscillation damping in applications involving parallel lines where fast control of series impedance is required.

Figure 11.12 depicts the electric diagram of a STATCOM used in a state farm application in Scotland.[6] Figure 11.13 illustrates the harmonic currents contributed to the harmonic spectrum measured during several hours at 33 kV connection point. Blue, red, and green bars represent maximum, average, and minimum values within the measurement period. The pink dashes are the planning levels in reference 7. Notice that the harmonic contribution from the STATCOM occurs in the range $h = 29$ to 43, and that they are well below the planning levels.

The static synchronous series compensator (SSSC) aimed at controlling reactive power flow through injection of voltage in series in an interconnected system uses four 12-pulse converters, which makes it a 48-pulse voltage source, with the 47th as the lowest harmonic, and hence with no need of harmonic filters given the low associated harmonic distortion levels. The DC bus of the SSSC allows for the provision of a substantial amount of energy storage. This can be achieved using a superconducting magnetic

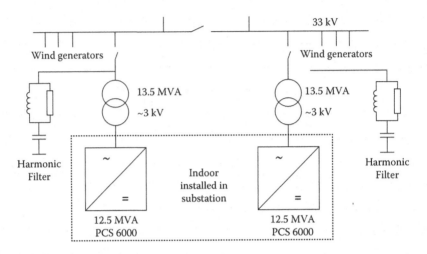

FIGURE 11.12
Example of a STATCOM installed in a wind park in Scotland. (Adapted from Maibach, P. et al., *STATCOM Technology for Wind Parks to Meet Grid Code Requirements*, European Wind Energy Conference (EWEC), Milan, Italy, 2007.)

energy storage (SMES) that can provide enhancement of transient stability in a multiarea system[8] by damping out power system oscillations, as depicted in Figure 11.14.

The superconducting coil is connected to the AC grid through two conventional 6-pulse converters arranged in a way that they form a 12-pulse

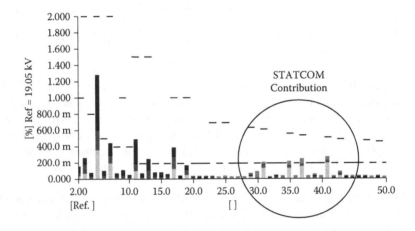

FIGURE 11.13
Harmonic currents measured at the connection point in a STATCOM application. (Adapted from Maibach, P. et al., *STATCOM Technology for Wind Parks to Meet Grid Code Requirements*, European Wind Energy Conference (EWEC), Milan, Italy, 2007.)

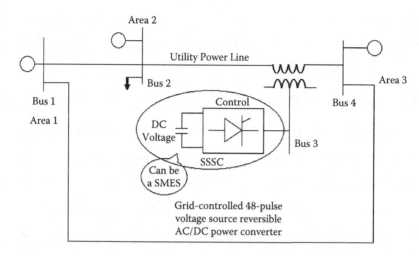

FIGURE 11.14
A single-phase diagram of an SSSC supported with a superconductive magnetic energy storage (SMES) device.

converter arrangement, as illustrated in Figure 11.15. Once charged with DC from the AC/DC conversion process, the SMES supports a magnetic field of approximately 1.2 T without any losses.[9]

Since a 12-pulse converter in the SSSC harmonic level has the 11th as the smallest harmonic, it should then be expected to operate with low and within recommended harmonic levels.

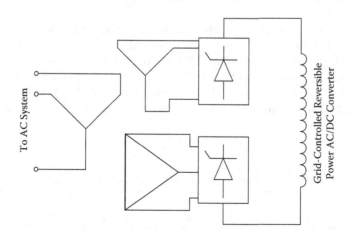

FIGURE 11.15
Diagram of a 12-pulse SMES.

TABLE 11.1

Example of AC Side Voltage Internal Requirements for the HVDC Light Version[10]

Individual harmonic distortion, $h \le 15$	$Dh = \dfrac{U_h}{U_1}$	$\le 1\%$
Individual harmonic distortion, $16 \le h$	$Dh = \dfrac{U_h}{U_1}$	$\le 0.5\%$
Total harmonic distortion	$THD = \sqrt{\displaystyle\sum_{h=2}^{50}\left(\dfrac{U_h}{U_1}\right)^2}$	$\le 1.5\%$
Telephone influence factor	$TIF = \sqrt{\displaystyle\sum_{h=2}^{50}\left(\dfrac{U_h TIF_h}{U_1}\right)^2}$	$\le 40\%$

Note: U_h is the hth harmonic (phase-to-ground) voltage, U_1 is the nominal fundamental frequency (phase-to-ground) voltage, and TIF_h is the weighting factor for the hth harmonic according to reference 11.

11.5 HVDC Harmonics and Filtering

Regarding harmonic levels, in addition to recommended limits by international standards, internal technical specifications of the particular HVDC version[10] have to be considered. Table 11.1 depicts the AC side voltage internal requirements stipulated for HVDC light version for individual and total harmonic distortion as well as for telephone influence factor.[11]

Further, worst case scenarios for short-circuit impedances have to be considered in determining individual harmonic distortion for the largest harmonic contributors.[11]

Likewise, the requirements for harmonic current specified in reference 10, depicted in Table 11.2, essentially follow IEEE 519 recommended limits, which were illustrated in Section 3.2.1 of Chapter 3.

DC filters, as depicted in Figure 11.16, are shunt connected in the electric system at the point where harmonic currents are generated and where they are meant to apply the filtering action much in the same way as AC filters

TABLE 11.2

AC Side Current Internal Requirements for the HVDC Light Version[10]

$\dfrac{I_{SC}}{i_L}$	$h < 11$	$11 \le h < 17$	$17 \le h < 23$	$23 \le h < 35$	$35 \le h$	THD
<50	2.00	1.00	0.75	0.30	0.15	2.50
≥50	3.00	1.50	1.15	0.45	0.22	3.75

Even harmonics are limited to 25% of the odd harmonic limits above.

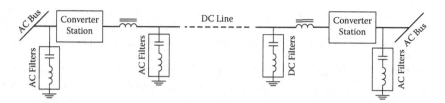

FIGURE 11.16
Block diagram of a 12-pulse SMES.

do. Harmonic DC filters are important in systems configured with a physical DC line to reduce interference with telephone subscribers, losses, overvoltages due to resonances, instability in firing pulse generation control, and interference with ripple control in load management applications.[12] Harmonic DC filters are usually smaller, and therefore more economical, than those on the AC sides. In back-to-back HVDC stations no DC filters are required.

The harmonics created at the DC side of the converter occur at multiples of the PWM pulse number p, expressed in terms of harmonic order h_dc as

$$h\_dc = (np), n = 1, 2, 3, \dots \tag{11.2}$$

References

1. Schlabbach, J., Harmonic Current Emission of Photovoltaic Installations under System Conditions, presented at the 5th International Conference on European Electricity Market 2008 (EEM 2008), May 2008.
2. Ahmed, A.L., Harmonic Impact of Photovoltaic Inverter Systems on Low and Medium Voltage Distribution Systems, MEng thesis, School of Electrical, Computer and Telecommunications Engineering, University of Wollongong, 2006.
3. Bollen, M., and Yang, K., Harmonics and Wind Power—A Forgotten Aspect of the Interaction between Wind-Power Installations and the Grid, *Elforsk Rapport*, 12, 51, August 2014.
4. Romero, A.A., Suvire, G.O., Zini, H.C., and Rattá, G., Possibilistic Harmonic Load-Flow for Electric Power Systems with Wind Farms, SICEL 2013, Medellin, Colombia, December 2013.
5. *ABB STATCOM for Flexibility in Power Systems*, ABB Pamphlet A02-0165E.
6. Maibach, P., Wernli, J., Jones, P., and Obad, M., *STATCOM Technology for Wind Parks to Meet Grid Code Requirements*, 3BHS237435, European Wind Energy Conference (EWEC), Milan, Italy, 2007.

7. *Planning Levels for Harmonic Voltage Distortion and the Connection of Non-Linear Equipment to Transmission Systems and Distribution Networks in the United Kingdom*, Engineering Recommendation G5/4, Energy Networks Association, London, February 2001.
8. Padma, S., and Lakshmipathi, R., Static Synchronous Series Compensator (SSSC) with Super Conducting Magnetic Energy Storage (SMES) for the Enhancement of Transient Stability in Multi Area System, *ACEEE Int. J. Control Syst. Instrum.*, 2(1), 2011.
9. Tripathy, S.C., Superconducting Magnetic Energy Storage, *Livewire*, 1(1), 2005.
10. Björklund, P.E., *HVDC Light Version 1*, Technical Specification 1JNL100088-337 Rev. 01, Internal Technical Specifications.
11. *The Telephone Influence Factor of Supply System Voltages and Currents*, EEI Publication 60-68, 1996.
12. Padiyar, K.R., *HVDC Power Transmission Systems*, New Age International Ltd. Publishers, New Delhi, 2005.

12

Harmonics from Latest Innovative Electric Grid Technologies

12.1 Introduction

Aging power systems infrastructure and increasing power demand constitute a serious challenge for electric power systems. Power system components, and particularly power circuit breakers, are witnessing an increased
fault current scenario often beyond their interrupting capabilities. This represents a problem not only for circuit breakers, but also for all other substation
equipment and hardware, given the large electrodynamic stresses developed during faults. Likewise, increased fault current levels mean higher step
and touch potentials during faults that may pose a security issue for some
substations.

Superconductive fault current limiters (SCFCLs) are under assessment by
utilities as a way to provide a low-loss device that can effectively reduce fault
currents to values comparable to prior fault current levels, much as if going
back in time and relaxing stresses in circuit breakers and substation equipment. This can allow for deferment of expensive switchgear equipment and
associated hardware replacement.

Diverse types of superconducting FCL devices are or have been under test,
among others, at Southern California Edison[1] and CE Electric at the UK[2] in
the last few years. Other types of SCFCLs include the solid-state and resistive types.[3] To some extent, all of these devices can potentially create some
sort of harmonic currents and voltage waveform distortion.

Other novel equipment with the potential to develop harmonic currents is the electric vehicle (EV) charging station. EV charging stations
are dramatically increasing in the world, principally motivated by the
idea to curb NO_2 emissions, which according to the International Energy
Agency,[4] together with fuel-stocked vehicles, will dramatically increase in
the next 20 to 30 years. This will certainly propel gas costs unless renewable resources come to the rescue.[4] Among the most feasible to see an
increase in CO_2, yet meeting expectations, are the battery electric vehicles,
plug-in hybrid electric vehicles, and fuel cell vehicles. What all this means

is an exponential growth that can amount to 27 million EVs sold by 2020 and over 1 billion by 2050.[4]

12.2 Electric Vehicles Connected to the Grid

Since all EVs use power electronics to get the batteries charged, in the long run this will certainly pose a challenge to electric utilities to keep the harmonic distortion levels within limits. This, particularly considering the potential for EVs to convert themselves into distributed generation sources, i.e., to feed power back to the grid, is a technology known as vehicle-to-grid (V2G).

This may allow utilities to use this concept to achieve some distribution management functions, including V2G vehicles charging from renewable resources connected to the power grid under light load conditions or supplying power to the grid to offer spinning reserves,[5] as illustrated in Figure 12.1.

Of course, all this will have to be carried out under compliance with international standards that regulate individual and total harmonic distortion in electric power systems[6] and in distributed resources connected to the grid.[7]

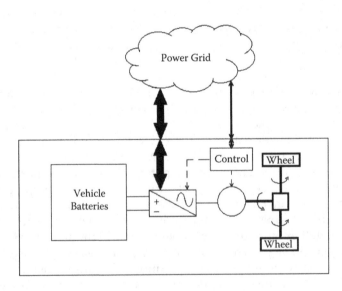

FIGURE 12.1
Vehicle-to-grid (V2G) concept.

Transmission line capacity has not kept pace with the generation capacity and load growth, with the result that transmission lines get subject to increasingly larger currents under fault conditions, stressing many components on the electric grid. Thus, fault current limiters (FCLs), innovative devices with the ability to protect the grid during fault conditions, have started to appear as an attractive option in the electric power utility to constrain the growth of fault current levels. The FCLs additionally offer to improve stability and quality of the entire electrical system.

However, at the same time, they can potentially increase existing levels of harmonic distortion. This short review refers to only one of the several fault current limiter technologies in the development of superconducting fault current limiters to illustrate that the extent of the harmonic distortion may be maintained well within limits as long as the operation of the SCFCL occurs within preestablished design parameters.

12.3 Superconducting Fault Current Limiters

Saturated-core fault current limiters use a permanent DC magnet to provide an operating condition in which the variation of flux vs. magnetic flux intensity in the *B–H* plane occurs in a region just above the knee of the curve, under normal operation conditions where a small slope on the *B–H* characteristic is observed, as illustrated in Figure 12.2.

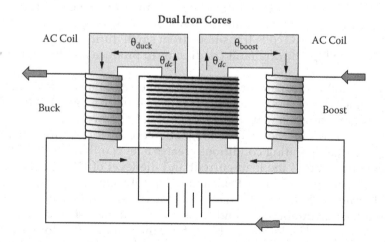

FIGURE 12.2
B–H characteristic of the SCFCL. (From IEEE 519-1992, *IEEE Recommended Practices and Requirements for Harmonic Control in Electrical Power Systems*. Reproduced with permission.)

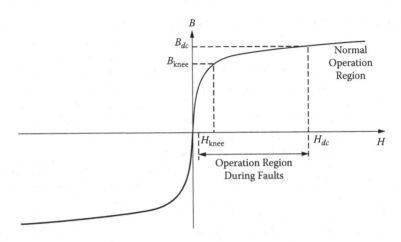

FIGURE 12.3 (See color insert.)
Simplified representation of a saturated-core SCFCL.

During a fault, the operation point on the B–H plane transitions from low (small slope in the B–H plane) to high (steep slope in the B–H characteristic), as depicted in Figure 12.2 for a single-phase SCFCL. The way this takes place is that the saturation of the magnetic flux provided by the DC magnet is influenced by the magnetic field of the two AC cores connected in series, but wound in different legs of the core, as shown in Figure 12.3.

At any particular time, the magnetic flux under one of the two AC coils is "boosted," while it is "bucked" at the portion of the magnetic core of the other AC coil. As the magnetic flux produced by the AC coils increases enough to drive the flux out of saturation, the impedance increases suddenly, acting to limit the AC line current flow. This is the way the changing impedance of the AC coils transitions from small during normal operation conditions to high during fault conditions, to produce a reduction of peak current.

The DC coil is a superconducting coil immersed in a cryogenic medium that produces the necessary amount of ampere turns to maintain the required level of core saturation with a small amount of DC power.

The operation of the SCFCL magnetic core in the saturation region under normal operating conditions creates a level of harmonic distortion that is a function of the load current and the level of DC bias applied to the magnetic core. Using a PSCAD model published in Moriconi et al.,[8] total harmonic distortion (THD) levels for voltage and current were calculated for maximum load conditions and maximum DC bias. The resultant calculated THD levels are 4.76% for voltage and 3.36% for current, as described in Figures 12.4 and 12.5, respectively, which are within recommended limits.[6]

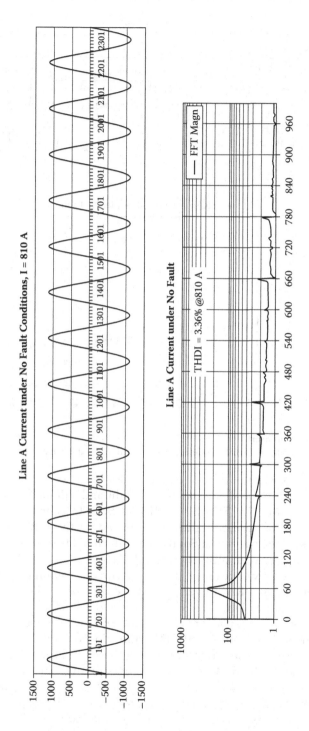

FIGURE 12.4
Line current and calculated THD for a saturated-core fault current limiter.

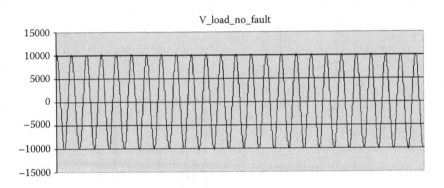

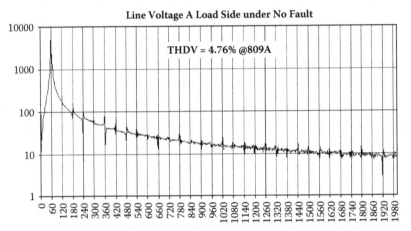

FIGURE 12.5 (See color insert.)
Line voltage and calculated THD for a saturated-core fault current limiter.

The harmonic spectrum of the voltage across the SCFCL calculated with the PSCAD program mentioned above for light and full load conditions is illustrated in Figure 12.6. For this particular SCFCL technology, the only documented case of harmonic distortion exceeding the IEEE 519[6] recommended limits refers to a case at Southern California Edison when a demonstrator SCFCL incurred an unlikely loss of DC bias.[9]

This was caused by a reset of the programmable automatic controller (PAC) and caused by a RAM overflow. At the time of the PAC problem, a resonance problem involving the SCFCL enlarged inductance and a nearby power factor correction capacitor bank occurred. This increased the voltage drop across the SCFCL, and as the SCFCL transitioned deeper into desaturation, odd harmonics started to increase until the level shown in Figure 12.7, obviously exceeding the IEEE 519 recommended limits, causing the voltage oscillations shown in Figure 12.8 and the automatic bypass of the SCFCL.[9]

The problem was clearly identified, analyzed, corrected, and properly documented, and the SCFCL was successfully reconnected thereafter.

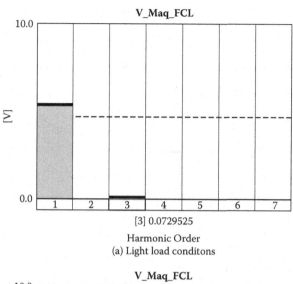

[3] 0.0729525

Harmonic Order

(a) Light load conditons

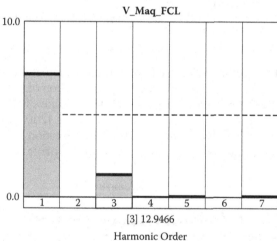

[3] 12.9466

Harmonic Order

(b) Full load conditons

FIGURE 12.6 (See color insert.)
Harmonic spectra of the calculated voltage across saturated-core fault current limiter for light and full load conditions.

12.3.1 Use of SCFCLs as a Means to Reduce Harmonic Sequence Currents during Faults, Leading to Wind Turbine Generator Transient Torque Reduction

The use of the saturated-core SCFCL to limit the SCC of different types of wind turbine generators (WTGs) was investigated in Muljadi et al.[10] Simulations were carried out to investigate the effectiveness of the FCL to

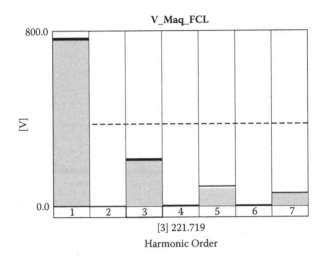

FIGURE 12.7 (See color insert.)
Harmonic spectra of the calculated voltage across saturated-core fault current limiter under a loss of DC bias implausible event.

limit the short-circuit current in a fault at the HV side of the transformer and to reduce transient torque during faults, via reduction of harmonic sequence currents at the time of fault. Figure 12.9 depicts one of the configurations studied, with a sketch of the WTG shown in Figure 12.10.

The result from this work revealed that the insertion of a saturated-core SCFCL provided the desired reduction of fault currents, as illustrated in

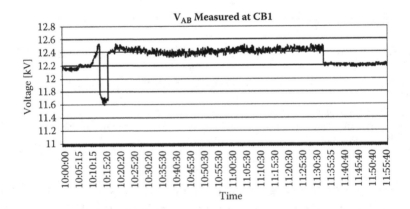

FIGURE 12.8 (See color insert.)
Line-to-line voltage in the SCE example at the time of the loss of DC bias in the SCFCL. (Adapted from Christopher, R.C. et al., Resonance of a Distribution Feeder with a Saturable Core Fault Current Limiter, presented at 2010 IEEE T&D Conference, New Orleans, April 19–22, 2010. With permission.)

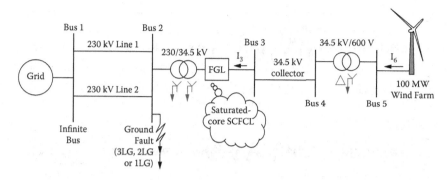

FIGURE 12.9 (See color insert.)
Saturated-core SCFCL as a means to reduced transient torque during faults. (Adapted from Muljadi, E. et al., Wind Power Plant Enhancement with a Fault Current Limiter, presented at Proceedings of 2011 IEEE Power and Energy Society General Meeting, Detroit, Michigan, July 24–29, 2011. With permission.)

Figure 12.11, and that it can significantly reduce the amplitude of positive-, negative-, and zero-sequence currents in the first few cycles during faults, as depicted in Figure 12.12.

It is important to mention that harmonic filters are usually considered for the control of harmonic distortion in wind farm applications when engineering assessment considering the specific type, configuration, and number of wind power turbines reveals that harmonic levels will be beyond recommended limits. By this, putting the saturated-core superconductive fault current limiter as a viable option to harmonic filters to offset harmonic sequence currents imposes undesirable transient torque during faults.

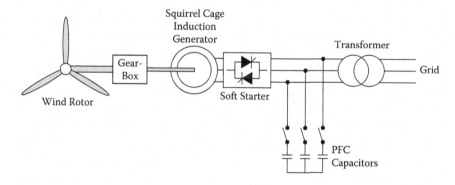

FIGURE 12.10 (See color insert.)
Wind farm with type I wind turbine induction generator. (Adapted from Muljadi, E. et al., Wind Power Plant Enhancement with a Fault Current Limiter, presented at Proceedings of 2011 IEEE Power and Energy Society General Meeting, Detroit, Michigan, July 24–29, 2011. With permission.)

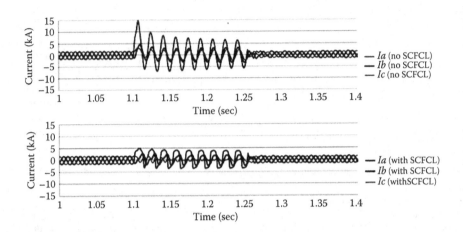

FIGURE 12.11 (See color insert.)
Single-phase-to-ground fault at the substation transformer. (Adapted from Muljadi, E. et al., Wind Power Plant Enhancement with a Fault Current Limiter, presented at Proceedings of 2011 IEEE Power and Energy Society General Meeting, Detroit, Michigan, July 24–29, 2011. With permission.)

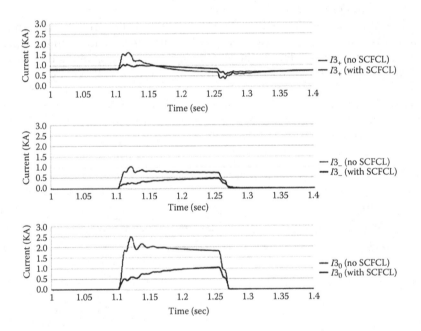

FIGURE 12.12 (See color insert.)
Sequence currents with and without SCFCL in the circuit. (Adapted from Muljadi, E. et al., Wind Power Plant Enhancement with a Fault Current Limiter, presented at Proceedings of 2011 IEEE Power and Energy Society General Meeting, Detroit, Michigan, July 24–29, 2011. With permission.)

12.4 Electric Vehicle Charging Stations

Charging stations for electric vehicles are no doubt a challenge for mid- and long-term power systems. The escalating increase of battery charges that is to be seen in the next decades indicates that without implementing a remedial measure, important power quality issues are to be expected in the power distribution grid as concurrent battery charging takes place.[11–14]

This relates to harmonic current distortion produced during current rectification processes.

Assessment for the potential for smart charging may involve third-party providers. The technical possibility of smart charging and how it may be possible to manage the EV charging so as to minimize the need for network reinforcement while delivering the needs of the EV user has to be further established.

State of charge and timing for concurrent battery charging are essential in the creation of the harmonic levels. Obviously, a severe scenario for harmonic distortion in a distribution feeder would be the case when all or a large percentage of EV chargers operate simultaneously. But even worse than that is when all those chargers are charged from zero or low charge state. However, the random nature of these processes may act so as to relieving the high harmonic levels that could potentially be reached under these critical situations, apart from the fact that different designs of battery charges, also described as diversity factor, may actually yield to some harmonic current cancellation.[11]

Battery chargers can represent additional losses and derating of transformers in a power system.[11,12] This is due to the additional heat produced by harmonic currents in the form of copper and core losses, as described in Chapter 9.

Ongoing work on the V2G has indicated that an EV with depleted batteries is nothing but a parasitic load when at the garage under charging cycles. As the car is plugged into a home outlet, it becomes connected to the electric network and its computer is responsible for interpreting if the car can begin to recharge its batteries or should help the power company during times of extreme consumption. Of course, an isolated case is insignificant, but in a city with thousands of V2Gs participating, important blocks of power could be produced to support the power utility supply.

V2G cars connected via the Internet-over-powerline connection system will be responsible for sending a signal from inside the cars' computer to a centralized server that handles the energy management of all network-connected cars, ready to deliver or absorb energy in terms of energy consumption in the electric system of the city. This system connecting to the network will act as an intermediary between the owner of the vehicle and the grid energy management companies, which are always trying to keep producing electricity at a constant level.[13]

When the network requires more energy due to increased demand, power companies usually draw this demand for traditional power plants, which are often coal, burning huge amounts of this mineral. When the vehicle network based on this cooperative V2G system achieves massive expansion, energy can be extracted from millions of vehicles plugged into their homes or garages, commercial fleets, such as the Postal Service vans, buses, trucks freight, and all those vehicles capable of providing energy support at times of peak consumption.

In essence, electric vehicles connected to the home network, and while stored in the garage, can be programmed to deliver their surplus energy to the grid, providing support to the power supply company during times of high energy consumption. It is during this period when the harmonic current contributions are expected to reach their peak.

It should be expected that in this scheme of power sharing using EVs to support electric companies and ultimately the community, over time the utility may end up paying the value of participating electric vehicles. This may only encourage increasing involvement from EV owners, which can represent a potential escalation in harmonic distortion levels.

V2G is still a new idea in the search for better ways to use energy, but is being closely pursued in the United States and Europe. After its onset in the last decade of the last century, it can now be considered a mature project with the potential to generate substantial amounts of energy in the short term. It is expected that this concept will be adopted mostly in countries that are implementing and strongly relying on renewable energy like sun and wind, such as Denmark and Britain.[13] It all points to the imminent embracing of this technology to meet our present needs, and this has been well understood by car manufacturers who are determinedly making V2G systems-based vehicles.

AC Propulsion of California[14] designed a V2G capable of producing 15 to 20 kW that could be enough to power around 7 to 10 houses in the United States while connected to the grid, just as any other distributed resource would do, but here in the form of a V2G energy storage supply. The assessment carried out by AC Propulsion revealed: "Integrating electric drive vehicles with the electric grid is technically practical and the concept has the potential to create an income stream that offsets a portion of vehicle ownership costs."[14] The vehicle demonstrated in that project was a pure battery EV with an 18 kWh battery.

Other automakers like Renault/Nissan, Mitsubishi, and BMW are producing all electric vehicles with an eye on the V2G market, which promises a massive launch at any time thanks to its system of cooperative operation.[14,15]

Figure 12.13 shows typical EV current harmonic spectra in two different battery charges measured in Malaysia.[16] Notice that the harmonic distortion as calculated by the measuring instrument shows a THD_I of 11.6 and 34.4%, respectively, for the modern EV charger and the golf cart battery charger. This study also found that modern EV chargers released lower THD_I than

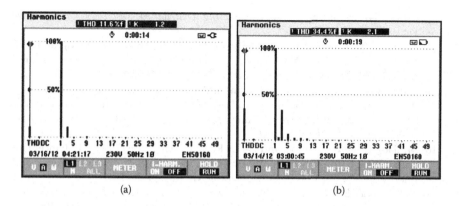

FIGURE 12.13 (See color insert.)
Typical current harmonic spectrum in (a) modern EV charger and (b) a golf cart charger. (From Wanic, M.Z.C. et al., *Engineering*, 5, 215–220, 2013. With permission.)

a golf cart charger, as it was expected, but that for THD$_V$, the modern EV caused a higher percentage of distortion than the golf cart charger. These results shed some light in the understanding of the harmonic distortion contribution from electric vehicle chargers, but the conclusions should not be generalized without taking into account any possible difference in the size of the batteries or in the charging rate.

References

1. Moriconi, F., De La Rosa, F., Darman, F., Nelson, A., and Masur, L., Development and Deployment of Saturated-Core Fault Current Limiters in Distribution and Transmission Substations, *IEEE Trans. Appl. Superconductivity*, 21(3), 1288–1293, 2011.
2. Klaus, D., Waller, C., Jones, D., McWilliam, J., Berry, J., Bock, J., Helm, J., Jafarnia, M., and Hobl, A., Superconductive Fault Current Limiters—UK Network Trials Live, in *CIRED, 22nd International Conference on Electricity Distribution*, Stockholm, June 10–13, 2013, Paper 0285.
3. Eckroad, S., Superconducting Fault Current Limiters, *Technology Watch 2009*, 1017793, technical update, December 2009.
4. A Look at the Electric Vehicle Movement, in *EV City Casebook*, 2012.
5. Kramer, W., Chakraborty, S., Kroposki, B., Hoke, A., Martin, G., and Markel, T., Grid Interconnection and Performance Testing Procedures for Vehicle-to-Grid (V2G) Power Electronics, presented at World Renewable Energy Forum 2012, Denver, CO, May 13–17, 2012.
6. IEEE 519-1992, *IEEE Recommended Practices and Requirements for Harmonic Control in Electrical Power Systems*.

7. IEEE 1547, *Standard for Interconnecting Distributed Resources with Electric Power Systems*.
8. Moriconi, F., Koshnick, N., De La Rosa, F., and Singh, A., Modeling and Test Validation of a 15kV 24MVA Superconducting Fault Current Limiter, presented at 2010 IEEE T&D Conference, New Orleans, April 19–22, 2010.
9. Christopher, R.C., Moriconi, F., Singh, A., Kamiab, A., Neal, R., Rodriguez, A., La Rosa, F., and Koshnick, N., Resonance of a Distribution Feeder with a Saturable Core Fault Current Limiter, presented at 2010 IEEE T&D Conference, New Orleans, April 19–22, 2010.
10. Muljadi, E., Gevorgian, V., and De La Rosa, F., Wind Power Plant Enhancement with a Fault Current Limiter, presented at Proceedings of 2011 IEEE Power and Energy Society General Meeting, Detroit, MI, July 24–29, 2011.
11. *The Potential Effects of Single-Phase Power Electronic-Based Loads on Power System Distortion and Losses: The Harmonic Impact of Electric Vehicle*, Vol. 4TR, 1000664, September 2003.
12. Staats, P.T., Grady, W.M., Arapaostathis, A., and Thallam, R.S., A Procedure for Derating a Substation Transformer in the Presence of Widespread Electric Vehicle Battery Charging, *IEEE Trans. Power Delivery*, 12(4), 1562–1568, 1997.
13. Letendre, S., and Kempton, W., The V2G Concept: A New Model for Power? Connecting Utility Infrastructure and Automobiles, presented at the 18th International Electric Vehicle Symposium and Exhibition, Berlin, Germany, October 20–24, 2001.
14. Gage, T., *Development and Evaluation of a Plug-In HEV with Vehicle-to-Grid Power Flow*, final report, CARB Grant ICAT 01-2, December 27, 2003.
15. Kempton, W., and Letendre, S., Electric Vehicles as a New Power Source for Electric Vehicles, *Transport. Res. D*, 2(3), 57–175, 1997.
16. Wanic, M.Z.C., Siam, M.F.M., Subiyanto, A.A., Mohamed, A., Azit, A.H., Mohamed, S.S., Mohamed, A., Hussein, Z.F., and Hussin, A.K.M., Harmonic Measurement and Analysis during Electric Vehicle Charging, *Engineering*, 5, 215–220, 2013.

Index

Printed in the United States
by Baker & Taylor Publisher Services